A Primer in Probability

STATISTICS: Textbooks and Monographs

Volume 1: The Generalized Jackknife Statistic, *H. L. Gray and W. R. Schucany*

Volume 2: Multivariate Analysis, *Anant M. Kshirsagar*

Volume 3: Statistics and Society, *Walter T. Federer*

Volume 4: Multivariate Analysis: A Selected and Abstracted Bibliography, 1957-1972, *Kocherlakota Subrahmaniam and Kathleen Subrahmaniam (out of print)*

Volume 5: Design of Experiments: A Realistic Approach, *Virgil L. Anderson and Robert A. McLean*

Volume 6: Statistical and Mathematical Aspects of Pollution Problems, *John W. Pratt*

Volume 7: Introduction to Probability and Statistics (in two parts) Part I: Probability; Part II: Statistics, *Narayan C. Giri*

Volume 8: Statistical Theory of the Analysis of Experimental Designs, *J. Ogawa*

Volume 9: Statistical Techniques in Simulation (in two parts), *Jack P. C. Kleijnen*

Volume 10: Data Quality Control and Editing, *Joseph I. Naus (out of print)*

Volume 11: Cost of Living Index Numbers: Practice, Precision, and Theory, *Kali S. Banerjee*

Volume 12: Weighing Designs: For Chemistry, Medicine, Economics, Operations Research, Statistics, *Kali S. Banerjee*

Volume 13: The Search for Oil: Some Statistical Methods and Techniques, *edited by D. B. Owen*

Volume 14: Sample Size Choice: Charts for Experiments with Linear Models, *Robert E. Odeh and Martin Fox*

Volume 15: Statistical Methods for Engineers and Scientists, *Robert M. Bethea, Benjamin S. Duran, and Thomas L. Boullion*

Volume 16: Statistical Quality Control Methods, *Irving W. Burr*

Volume 17: On the History of Statistics and Probability, *edited by D. B. Owen*

Volume 18: Econometrics, *Peter Schmidt*

Volume 19: Sufficient Statistics: Selected Contributions, *Vasant S. Huzurbazar (edited by Anant M. Kshirsagar)*

Volume 20: Handbook of Statistical Distributions, *Jagdish K. Patel, C. H. Kapadia, and D. B. Owen*

Volume 21: Case Studies in Sample Design, *A. C. Rosander*

Volume 22: Pocket Book of Statistical Tables, *compiled by R. E. Odeh, D. B. Owen, Z. W. Birnbaum, and L. Fisher*

Volume 23: The Information in Contingency Tables, *D. V. Gokhale and Solomon Kullback*

Volume 24: Statistical Analysis of Reliability and Life-Testing Models: Theory and Methods, *Lee J. Bain*

Volume 25: Elementary Statistical Quality Control, *Irving W. Burr*

Volume 26: An Introduction to Probability and Statistics Using BASIC, *Richard A. Groeneveld*

Volume 27: Basic Applied Statistics, *B. L. Raktoe and J. J. Hubert*

Volume 28: A Primer in Probability, *Kathleen Subrahmaniam*

OTHER VOLUMES IN PREPARATION

A Primer in Probability

KATHLEEN SUBRAHMANIAM
Department of Statistics
University of Manitoba
Winnipeg, Manitoba, Canada

MARCEL DEKKER, INC. New York and Basel

Library of Congress Cataloging in Publication Data

Subrahmaniam, Kathleen.
A primer in probability.

(Statistics, textbooks and monographs ; v. 28)
Includes index
1. Probabilities. I. Title.
QA273.S766 519.2 79-12677
ISBN 0-8247-6836-1

Marcel Dekker, Inc.
270 Madison Avenue, New York, New York 10016

Current printing (last digit):
10 9 8 7 6 5 4

Printed in the United States of America

To my husband

Contents

Preface

A Primer in Probability is an outgrowth of a first year university course that the author has been teaching for several years. The *Primer* presents elementary probability theory as a basic building block for statistical inferences and its applications. It is envisaged that a course based on this text can be taught in either a semester or a quarter. As preparation, students are expected to have only a thorough background in high school algebra.

The *Primer* can be used to provide two types of courses: (1) a course for mathematical science students interested in detailed proofs, (2) a less mathematically oriented course emphasizing the results of the proofs and examples. This second type of course might well be offered for high school seniors.

At this level a course in probability should be oriented toward "problem solving" for only by working problems do students really master the subject. Worked examples, taken from a variety of disciplines, are provided throughout the text. To encourage students to work more problems, solutions, some in considerable detail, are given for the exercises at the end of each chapter.

As it is actually the students who are the real critics of any text, the author is indeed grateful to the many students who have provided helpful suggestions concerning the text and the problems.

It is hoped that the layout of the manuscript will make the book more readable. The expert typing and retyping to specification have been provided by Mrs. Eva Loewen.

Certainly at the inception of this project, the author never realized the many hours of work that would be needed for its completion. The author wishes to thank her son Narayana for going through the manuscript and providing a better perspective on the problems in the book. To her son Rama, the author is equally thankful for his writing the programs for the HP65 used in computing the tables included in the text. To her husband, the author is most grateful for his endless encouragement and thought-provoking discussions. Without his help, this *Primer* would never have been a reality.

Finally the author wishes to thank Dr. D. B. Owen for bringing this book to the attention of the publishers.

Kathleen Subrahmaniam

A Primer in Probability

Chapter 1

A First Glimpse of Probability

1-1 What Is Probability?

It is not uncommon to hear the term *probability* used in day-to-day conversation. Two friends, planning a golf game tomorrow, may ponder the question: What is the probability of rain tomorrow? A young man of 25, having just paid the premium for his retirement insurance, may question as to the probability of his living to age 65 so that he may realize its benefits. After just paying an enormous medical bill for the correction of a slipped disc, a patient enquires of his doctor: "What is the probability that I will again need this kind of operation?"

In each of these situations the term *probability* carries with it an idea or concept of the *chance* or *likelihood* of an event.

How did the idea of probability as a measure of chance arise? In the seventeenth century, games of chance were very popular, particularly in France. In an attempt to improve his fortune, the Chevalier de Méré, an avid French gambler with a passing interest in mathematics, consulted the eminent mathematician and theologian Blaise Pascal.

Among the problems which de Méré posed to Pascal was the "problem of points." In the "problem of points," two people (I and II) play a set of games in which each has an equal opportunity of winning a point. The first player to score 5 points wins the set. If the set is prematurely terminated with player I having 4 points and the other having 3 points, how should the stakes be divided?

One might argue that the stakes should be divided 4 to 3. Pascal, however, said the stakes should be split 3 to 1 in favor of the person having 4 points. Do you think that this 3 to 1 split is reasonable? If two more games were played the set would have to be completed:

Outcome	*Winner of Set*
I wins, I wins	I
I wins, II wins	I
II wins, I wins	I
II wins, II wins	II

Player I wins three times out of four!

Another of de Méré's enquiries involved the tossing of dice, He had learned, possibly from experience, that a balanced six-sided die must be tossed four times in order that the chance of getting at least one six will be greater than one half. From this he argued that it should be advantageous to bet on the occurrence of at least one double-six in 24 tosses of a pair of dice. When he lost his bet, he complained to Pascal. Pascal assured him that these results were to be expected: the probability of at least one six in four tosses is $1 - (5/6)^4 = .518$, but the probability of at least one double-six in 24 tosses of a pair of dice is $1 - (35/36)^{24} = .491$.

Motivated by these enquiries, Pascal engaged in a lengthy correspondence with another mathematician, Fermat. This correspondence served as a basis for the unified theory of chance phenomena which we today call the theory of probability. Although the initial interest in the problem was sparked by gambling, investigators during the past three centuries have noted the analogy between laws of uncertainty involved in games of chance and the laws of variation observed in apparently uncontrollable phenomena.

What do we mean by uncontrollable phenomena? Let us consider two examples: (1) tossing a balanced coin and noting the frequency of heads, (2) noting the frequency of males in human birth.

Table 1·1 relates to the proportion of males in sequences of human births. The results are given for 10 sequences of 10, 50 and 250 births. How else might this type of data arise? A similar table might have resulted from recording the proportion of heads when a balanced coin is tossed; that is, the pattern of variability in the two examples is similar. In fact, we might conclude that the *probability of a male birth* as well as the *probability of a head* is equal to .5.

Table 1·1 Proportion of Male Births

Sequence Number	*Proportion of Males in*		
	10 Births	*50 Births*	*250 Births*
1	.5	.47	.512
2	.5	.50	.524
3	.4	.51	.484
4	.3	.60	.500
5	.4	.52	.504
6	.6	.49	.476
7	.6	.50	.508
8	.5	.51	.472
9	.9	.53	.516
10	.2	.46	.496

In these situations we see that certain events do not happen in a clearly predictable or deterministic fashion. For example, tossing a balanced coin may result in either a "head" or a "tail"; however, before the coin is tossed we do not know the outcome. Similarly, prior to birth the sex of the child is not known.

In order to understand the role of probability in interpreting this type of uncertain outcome, let us investigate the properties of deterministic and random experiments.

1-2 Experiments: Deterministic or Random

What do we mean by an experiment? What is the role of probability in interpreting the results of an experiment?

In the chemistry laboratory the students determine the boiling point of water as 100°C. Any deviation from this is a result of changing experimental conditions such as elevation of pressure or experimental error introduced by failure to read the thermometer correctly. If the experimental conditions remain the same, then the determination will be the same.

In contrast there are experiments in which the results vary in spite of all efforts to keep the experimental conditions the same. For example, coin tossing or the birth data can be thought of as experiments". In these experiments the results are unpredictable.

As we have seen from our previous examples, not all experiments must necessarily take place in a scientific laboratory. In general we shall use the word experiment to describe any act that can be repeated under given conditions. If the results of the repeated experiments are exactly the same, we say that the experiments are *deterministic*; otherwise, they are said to be *random* or *stochastic*.

We shall see that probability theory is used to explain and predict, to some degree, the results of random experiments.

1-3 The Role of Probability in Statistical Inference

Statistics, as a discipline, is the science of the collection and analysis of data with a view to drawing inferences about populations. When data are gathered, we may use statistical inference to choose among alternative models. The drawing of inferences relies to a great extent on probability theory.

Consider selecting a sample of voters to estimate the proportion of voters favoring a particular candidate. If we know the proportion of individuals in the population who favor the candidate, we can use probability theory to predict what the fraction in the sample is likely to be. This, however, is not the problem with which we are usually faced. In contrast we determine the proportion in the sample and wish to draw conclusions about the proportion in the population. Probability theory deduces from the known content of the population the probable content of the sample; on the other hand, statistical inference is a means of describing the content of the population from the observed sample.

In the example we have been discussing, the *population* is a large (countable) number of people and the *sample* represents a subset of the people. In contrast to this the population may only be hypothetical or conceptual. As an illustration consider the coin-tossing experiment. Here the population consists of all the possible times the coin might be tossed--surely a situation which can only be hypothetical. Every sequence of tosses represents a "sample" from the conceptual population.

In summary, then, we have

1. Population (known) → sample (unknown): Deductive reasoning or arguing from the general to the specific.
2. Population (unknown) ← sample (known): Inductive reasoning or arguing from the specific to the general.

Deduction answers such questions as: With a given population, how will the sample behave? Will the sample represent the population? We shall see that only when this deductive problem is resolved can we turn around this argument and ask: How precisely can we describe an unknown population characteristic based on an observed sample?

1-4 Interpreting Probability

1-4-1 Empirical Basis of Probability. We have seen that in random experiments the outcome for a particular trial may be unpredictable. If, however, we examine repetitions of a random experiment, we note that a pattern emerges as the number of repetitions N increases. We have already seen evidence of this in Table 1·1.

As a further illustration let us suppose that we are conducting a telephone survey to investigate whether or not people are watching the television between five and six o'clock in the evening. Each of 10 interviewers makes 10 telephone calls. The data in Table 1·2 are the results of such a survey.

Table 1·2 Telephone Survey of TV Audience

Interviewer	*Relative Frequency in 10 Calls*	*Cumulative Relative Frequency*
1	.3	3/10 = .300
2	.4	7/20 = .350
3	.2	9/30 = .300
4	.2	11/40 = .275
5	.6	17/50 = .340
6	.5	22/60 = .367
7	.1	23/70 = .329
8	.3	26/80 = .325
9	.2	28/90 = .311
10	.2	30/100 = .300

The interviewer will find that the *relative frequency*:

$$\text{r.f.} = \frac{\text{number of households in which television is being watched}}{\text{number of households called}}$$

may vary from 0 to 1. If, however, a large number of interviewers each repeats this experiment (or a single interviewer repeats the experiment a large number of times), the relative frequency seems to become stabilized and approach a constant value. We say that the relative frequency is converging to its probability. Although the relative frequencies obtained by individual interviewers may vary considerably, we see that as we successively combine the results of the interviewers the overall relative frequency seems to approach .3. From this we would conclude that the probability of an individual watching television in the hour from five to six o'clock is .3.

The frequency approach to probability defines the probability of an event E as

$$\Pr(E) = \lim_{n\to\infty} \frac{n_E}{n}$$

where n_E is the number of times the event E occurs in n trials.

The data in Table 1·1 illustrate the concept of "long-run stability" upon which this empirical approach to probability is based. Although the result on any particular trial of the experiment cannot be predicted, a long sequence of trials reveals a pattern in which the relative frequency or probability tends to stabilize.

1-4-2 Classical Definition. As we have mentioned, much of the early development of probability theory was motivated by games of chance in which dice and cards were used. The most familiar gambling device is a die--a cube with six symmetrical faces. Most of us, even without any experimentation, would feel that each face of the cube is *equally likely* to appear when the die is rolled. If each of the six outcomes is equally likely, then the probability associated with each face is 1/6.

In the classical definition of probability each of the k equally likely possible outcomes is assigned a probability of 1/k. Then the probability of an event E is defined as the ratio of m, the number of outcomes favoring E, to k, the total number of equally likely outcomes:

$$\Pr(E) = \frac{m}{k} .$$

As we shall see, there are several pitfalls in this definition:

1. It is logically circular in that it defines *probability* in terms of *equally likely* outcomes. Since we are defining probability, the phrase *equally likely* or *with equal probability* has not been defined.

2. As this definition is applicable only to equally likely outcomes, what would we do if the die had been weighted so that even numbers were twice as likely as odd ones?

In practice we shall see that the assignment of probability using the classical definition is often reinforced by experimentation; that is, the classical and frequency approaches are combined.

1-4-3 Subjective Probability. In addition to the equiprobable and frequency approaches to probability, we might also consider a subjective or personal interpretation of probability.

How probable is rain in the afternoon when it is cloudy in the morning? In the answer to this question, probability is viewed as a measure of personal belief. This measure of belief may vary from individual to individual even when they are confronted with the same set of evidence. Given a cloudy morning sky, you may well prepare for rain, but your friend may not.

It may be somewhat comforting to note that based on a large amount of data, the frequencist and subjectivist will usually agree on the assignments of probability.

Problems

1. A simple experiment consists of tossing three balanced coins simultaneously. Consider the four outcomes:

   A: no head occurs

   B: exactly one head occurs

   C: exactly two heads occur

   D: exactly three heads occur.

   Two friends, Don and Harry, have differing ideas about the probabilities of these outcomes. Don thinks that the outcomes A, B, C and D should have the probabilities 1/4, 1/4, 1/4, 1/4, respectively, whereas Harry thinks the appropriate probabilities are 1/8, 3/8, 3/8, 1/8.

(a) Perform the above experiment 32 times and record the number of occurrences of each of the outcomes A, B, C and D.

(b) Which of these two assignments of probabilities seems more plausible in light of the data?

(c) If you had repeated the experiment 100 times, what do you think would have happened?

2. Open the telephone book to any page and obtain the frequency distribution of the last digit for 25 numbers.

(a) Find the average value of the last digit and compare it with 4.5, the theoretical value when all digits are equally likely. Why does the observed average deviate from the theoretical one?

(b) If you choose 100 numbers rather that 25, what would you expect to happen? Why?

3. Distinguish between deterministic and random experiments. Give an example of each.

4. Suppose a man stands facing north and tosses a coin to decide to walk to the north one step or south one step. He continues tossing and then walking for 25 steps. We call this a "random walk."

(a) Without any experimentation

(i) how many steps do you think he will be on the north side of the starting point?

(ii) how many steps will he be on the south side of the starting point?

(iii) how many times will he have returned to the origin?

(b) Now simulate the experiment by tossing a coin and moving a point one step north or south, corresponding to a head or a tail, respectively. For example, if four tosses result in HTHH, then the random walk ends two steps north of the starting point. The walker has been north of the origin for three steps and once he has returned to the starting point. With this sequence he has never been south of the starting point. Toss the coin 25 times and answer the questions in (a).

(c) Are the results in (b) what you would have expected?

(d) Repeat the "random walk" five times. How does this added experimentation affect your answer?

5. Refer to the problem of points in Section 1-1. Using a balanced coin simulate the finish of the set 20 times. Compare the number of times player I wins with the number of times player II wins. Are these numbers approximately in the ratio 3 to 1 as Pascal predicted?

6. Suppose Jack and John are playing a game in which each has an equal opportunity of winning. At the start of the game, Jack has $3 and John has $2. The boys stake $1 on each game. If they play until one of them is ruined, the probability that Jack wins should be approximately .6.

   (a) How would you simulate a single game using a balanced coin?

   (b) Using simulation, play 10 games and determine the relative frequency of the games which Jack wins. Comment on this result.

Chapter 2

Basic Concepts of Probability

2-1 Sample Space

In our discussion of random experiments we have seen that there may be many outcomes of an experiment. Due to the chance mechanism involved, the exact outcome of a random experiment cannot be predicted with certainty. How shall we describe these outcomes in a more structured way? We shall use the concepts of set theory developed in Appendix B for developing a formal basis for probability.

Think of the simple experiment of tossing a coin. We would generally agree that a head or a tail are the only possible outcomes. If we denote these outcomes by H and T, then each possible outcome of the experiment would correcpond to exactly one of the elements in the set {H, T}. This set of outcomes is called a sample space for the experiment and we write S = {H, T}.

Let us now consider a slightly more complicated experiment: Suppose we toss simultaneously two distinguishable coins, a penny and a nickel. How shall we record the possible outcomes of this experiment? Clearly the outcome for each coin is either a head or a tail. We could record only the total number of heads on both coins. Then $S_1 = \{0, 1, 2\}$ corresponds to _a_ sample space. This, however, is not the only way of describing the experiment. For instance, this does not allow us to answer the question: Did the penny fall heads? To answer this question, we need a sample space with a finer classification: $S_2 = \{H_PH_N, H_PT_N, T_PH_N, T_PT_N\}$. Here the symbol H_PH_N denotes a head on both coins, while H_PT_N means a head on the penny and a tail on the nickel. We have encountered a typical situation in which there is no one correct or unique sample space for the experiment. Different people, or even the same person at a different time, may describe the outcomes differently. The characterization of the outcomes depends upon the questions to be answered. We shall soon see that it is in general a safe guide to include as much detail as possible in the description of the outcomes.

With these examples in mind, we will now formally define the term *sample space* using the mathematical terminology of set theory.

> Definition 2·1: *A* sample space *of an experiment is a set S of elements* $E_1, E_2, \ldots, E_k$ *such that any outcome of the experiment corresponds to exactly one element in the set. The elements* $E_1, E_2, \ldots, E_k$ *are called* sample points.

2-2 Events and Their Probabilities

In this section we shall further extend the analogy between the principles underlying the description of a random experiment and set theory.

With reference to the example concerning the tossing of a penny and nickel, we may be interested in particular outcomes which we will label events:

U:	Exactly one head appears	$U = \{H_PT_N, T_PH_N\}$
V:	Exactly two heads appear	$V = \{H_PH_N\}$
W:	At least one head appears	$W = \{H_PT_N, T_PH_N, H_PH_N\}$
X:	A head appears on the penny	$X = \{H_PH_N, H_PT_N\}$
Y:	A head appears on the nickel	$Y = \{T_PH_N, H_PH_N\}$
Z:	No head appears	$Z = \{T_PT_N\}$

What property do these sets have in common? Reference to the sample space $S_2 = \{H_P H_N, H_P T_N, T_P H_N, T_P T_N\}$ shows that these sets U, V, W, X, Y and Z are each a subset of S_2.

Definition 2·2: *An* event *is a subset of the sample space* S.

We will say that the event E has occurred if the outcome of the experiment corresponds to an element in the subset E.

When we consider these six events, we see that V and Z are different from the others in that each contains only one sample point. Events containing exactly one sample point will be called *simple events*. Events made up of more than one sample point will be called *composite*. Note that composite events can always be decomposed into simple events.

Since we are concerned with random experiments, just a list of the outcomes will not fully describe the experimental set-up. We have seen that each of these outcomes has a certain probability or likelihood associated with it. Suppose we have the sample space

$$S = \{E_1, E_2, \ldots, E_k\}.$$

How shall we assign probabilities to these outcomes? That is, we wish to determine

$$\Pr(E_i) \quad \text{for } i = 1, 2, \ldots, k.$$

In the previous chapter we indicated three ways in which these probabilities might be assigned:

1. We may feel intuitively, or supported by experimentation, that each of the k outcomes is equally likely. Then

$$\Pr(E_i) = \frac{1}{k} \quad \text{for } i = 1, 2, \ldots, k.$$

2. In other situations we may find it convenient to rely on the observed relative frequencies obtained from a series of repeated experiments.

$$\Pr(E_i) = \text{relative frequency of the event } E_i.$$

3. The probabilities may be assigned from a personal point of view. In this case we say the probabilities are subjective.

In each of these three cases we have assigned the probabilities so that

$$0 \leq Pr(E_i) \leq 1$$

and

$$\sum_{i=1}^{k} Pr(E_i) = 1.$$

Combining the enumeration of the possible outcomes and their assigned probabilities, we can define a probability model.

> Definition 2·3: *The sample space* $S = \{E_1, E_2, \ldots, E_k\}$ *and the assigned probabilities* $Pr(E_i)$ *for* $i = 1, 2, \ldots, k$ *determine the* probability model *for a random experiment.*

From the probability model we can obtain the probability of any event associated with the experiment.

> Definition 2·4: *The* probability of any event E *is the sum of the probabilities of the simple events which constitute the event* E.

Two somewhat special cases arise: the entire sample space and the impossible event. Since all the possible outcomes of an experiment must be enumerated in the sample space, $Pr(S) = 1$. This would be translated as "some event in S must occur," which seems very reasonable from the definition of S. If any event is not a possible outcome of the experiment, then it has no corresponding sample points in S. We will call this event an "impossible event" and its probability is obviously zero.

Referring to our penny-nickel experiment, let us develop an appropriate probability model and calculate the probabilities corresponding to the events U, V, W, X, Y and Z. It would seem reasonable, and it can be verified by experimentation, that each outcome is equally likely. Assigning probability 1/4 to each point in S_2 and using Definition 2·4, we find that Pr(U) = 1/2, Pr(V) = 1/4, Pr(W) = 3/4, Pr(X) = 1/2, Pr(Y) = 1/2 and Pr(Z) = 1/4.

2-3 Combining Events

Since sets and events are analogous, we will now discuss how events, like sets, may be combined. In combining events we are faced with the problem of translating "words" into logical expressions. In every day usage, expressions of the form "A or B" may be interpreted in two different ways:

1. Exclusive "A or B but not both"
2. Inclusive "A or B or both."

In the following discussion we shall restrict ourselves to the inclusive form.

Definition 2·5: *The* union *of the events* A *and* B *in* S *is the set of all points that belong to at least one of the sets* A *and* B.

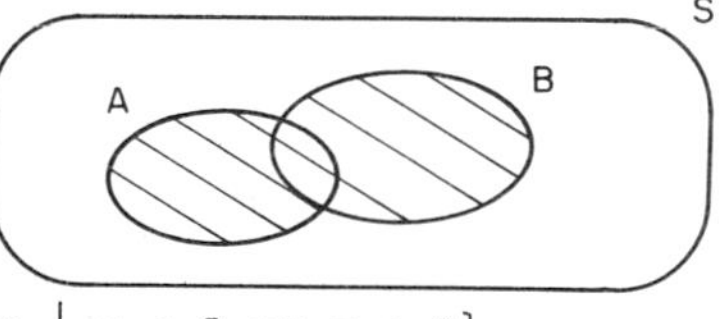

$A \text{ or } B = A \cup B = \{x \mid x \in A \text{ or } x \in B\}.$

Simultaneous membership in two sets A and B is expressed in words by the terminology "and."

Definition 2·6: *The* intersection *of the events* A *and* B *in* S *is the set of all points belonging to* A *and to* B.

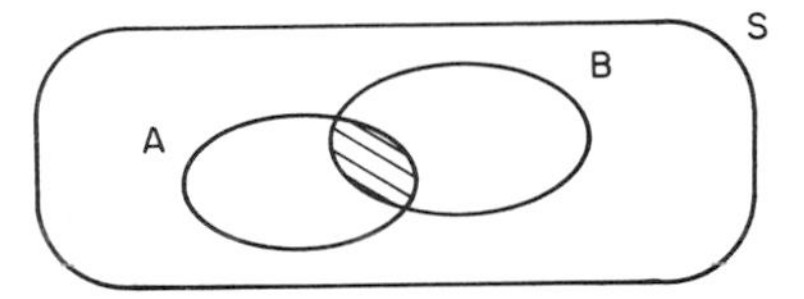

A *and* B = AB = $A \cap B = \{x \mid x \in A$ *and* $x \in B\}$.

Two events which cannot occur at the same time are said to be *mutually exclusive* (m.e.). We would speak of the sets corresponding to these events as being disjoint since they have no points in common. We use the symbol ϕ to designate the null set.

Definition 2·7: *If* $A \cap B = \phi$, *then the events* A *and* B *are mutually exclusive*.

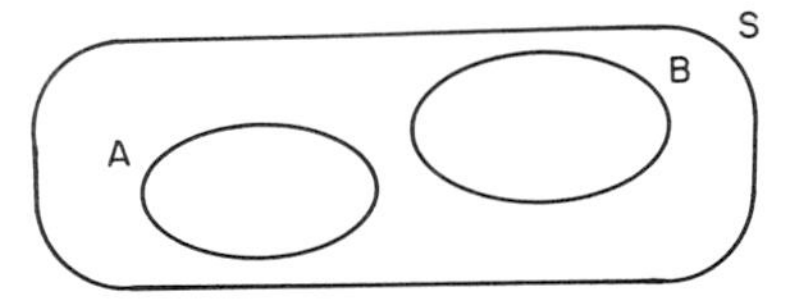

$A \cap B = \phi$.

If the events $A_1, A_2, \ldots, A_k$ are mutually exclusive (i.e., the intersection of any pair is ϕ) and exhaustive (i.e., their union is S), then we say that these k events form a *partition* of the sample space. Note that the simple events which describe the sample space always form a partition of S.

Often we may be interested in the fact that a particular event has not occurred.

> Definition 2·8: *The event* $\bar{A}$, *consisting of all the points in* S *which are not contained in* A, *is called the* <u>complement</u> *of* A.
>
> 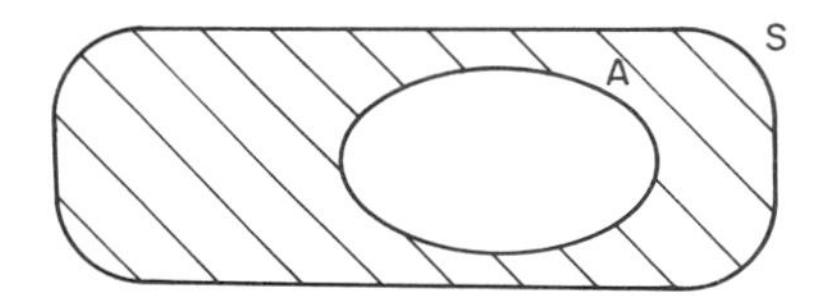
>
>
> $\bar{A} = \{x \mid x \notin A\}$.

In passing we note that A and $\bar{A}$ are m.e. and form a partition of S since $A \cup \bar{A} = S$.

In discussing events it will often be helpful to decompose an event into the union of m.e. components. Suppose the events $A_1, A_2, \ldots, A_k$ form a partition of S. Then any event F in S can be written as

$$F = (A_1 \cap F) \cup (A_2 \cap F) \cup \cdots \cup (A_k \cap F).$$

To illustrate this concept, consider the following Venn diagram in which B_1, B_2, B_3 form a partition of S.

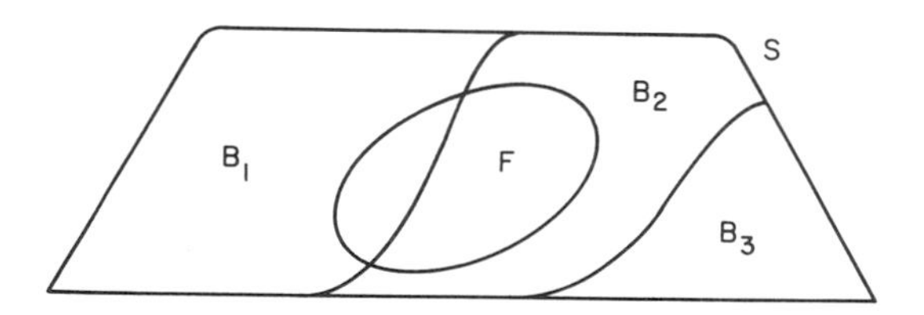

$$F = (B_1 \cap F) \cup (B_2 \cap F) \cup (B_3 \cap F).$$

Note that in this case the events B_3 and F have no points in common; thus $B_3 \cap F = \phi$. Now consider the special case of the event B_1 and its complement $\bar{B}_1$. Since B_1 and $\bar{B}_1$ form a partition of S, we can write

$$F = (B_1 \cap F) \cup (\bar{B}_1 \cap F).$$

In the above drawing, $\bar{B}_1 = B_2 \cup B_3$.

To illustrate this result, consider a bolt-manufacturing plant. The bolts are produced in three shifts: 8 to 4, 4 to 12 and 12 to 8. Since the manufacturer is interested in maintaining the quality of production, he wishes to monitor the defective bolts being produced. The defectives can be produced on any one of the three shifts. Let Sh_1 represent the 8 to 4 shift, Sh_2 the 4 to 12 shift and Sh_3 the 12 to 8 shift. Since Sh_1, Sh_2 and Sh_3 are m.e., the event D, production of a defective bolt, can be written as

$$D = (Sh_1 \cap D) \cup (Sh_2 \cap D) \cup (Sh_3 \cap D).$$

Another useful combination of sets will be the *difference* of two sets.

Definition 2·9: *The* difference *of* A *and* B, *or the* relative complement *of* B *with respect to* A, *is the set of all points in* S *which belong to* A *but not to* B.

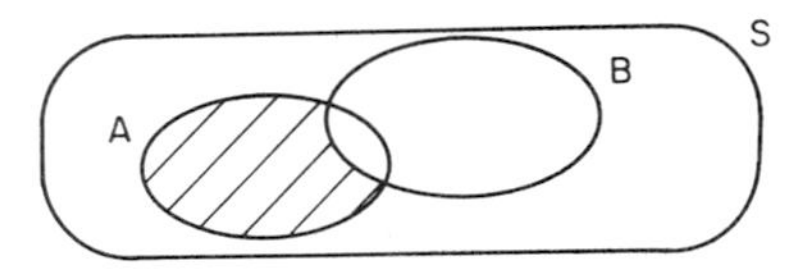

$$A - B = A \cap \bar{B} = A\bar{B} = \{x \mid x \in A \text{ and } x \notin B\}.$$

2-4 Probabilities Associated with Combined Events

From our discussion of the combination of events, we have seen that the combined events are new events and hence subsets of S. How shall we determine the probability of these combined events? In Section 2-2 we found the probability of events by a direct sample-point approach. In many situations we will find that this may be a difficult task. In this section we will develop some rules which will often simplify the determination of probabilities. Referring to our Definition 2·4 of the probability of an event, it would seem reasonable to say that if A and B are m.e. events, then the probability of their union $(A \cup B)$ is just the sum of their individual probabilities.

In our penny-nickel example we found $Pr(W) = 3/4$. Now the event W can be written as $U \cup V$ where U and V are m.e.; hence, $Pr(W) = Pr(U) + Pr(V) = 1/2 + 1/4 = 3/4$.

The rationale underlying the laws of probability which we will develop may easily be explored using Venn diagrams. Our approach will be somewhat more rigorous in that we will develop the laws starting with the following three axioms:

Axiom 1: $0 \leq Pr(A) \leq 1$

Axiom 2: $Pr(S) = 1$

Axiom 3: $Pr(A_1 \cup A_2) = Pr(A_1) + Pr(A_2)$, where A_1 and A_2 are m.e.

Addition Law: *If* A *and* B *are any two events in* S, *then the probability of their union is*

$$Pr(A \cup B) = Pr(A) + Pr(B) - Pr(A \cap B).$$

Proof:

(i) From Definition 2·4 we know that the probability of $A \cup B$ is the sum of the probabilities of the points in $A \cup B$. By inspection of the Venn diagram of $A \cup B$, we see that $Pr(A) + Pr(B)$ is the sum of the probabilities of the points in A and of the probabilities of the points in B. The portion in the intersection $(A \cap B)$ has been counted twice, once in A and once in B. Correcting for the double counting, we have

$$Pr(A \cup B) = Pr(A) + Pr(B) - Pr(A \cap B).$$

(ii) We will now prove the addition law more formally using the three basic axioms of probability. Since our axioms involve m.e. events, we will find it useful for the purpose of proof to decompose the event of interest into its m.e. components. The event $A \cup B$ can be written as $A \cup (\bar{A} \cap B)$, where A and $(\bar{A} \cap B)$ are m.e.

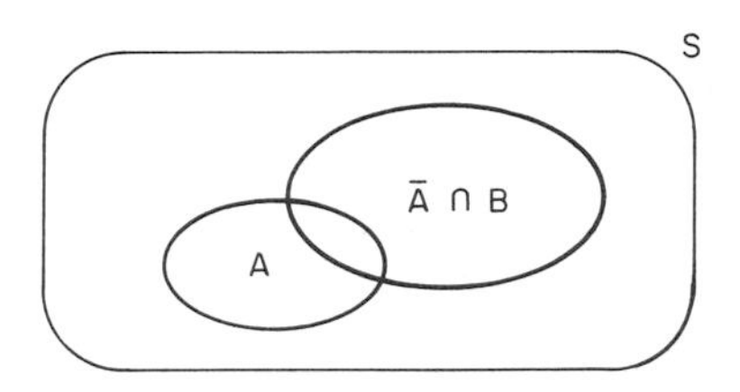

Using Axiom 3, we then have

$$\Pr(A \cup B) = \Pr(A) + \Pr(\bar{A} \cap B).$$

The event B can also be written as the union of m.e. events: $B = (A \cap B) \cup (\bar{A} \cap B)$. Hence,

$$\Pr(B) = \Pr(A \cap B) + \Pr(\bar{A} \cap B)$$

or

$$\Pr(\bar{A} \cap B) = \Pr(B) - \Pr(A \cap B).$$

Substituting into the expression for $\Pr(A \cup B)$, we obtain

$$\Pr(A \cup B) = \Pr(A) + \Pr(B) - \Pr(A \cap B). \qquad \square$$

In our example consider the event of a head on the nickel or the penny; that is, $X \cup Y$. Applying the addition law, we have

$$\Pr(X \cup Y) = \Pr(X) + \Pr(Y) - \Pr(X \cap Y).$$

Noting that $(X \cap Y)$ equals V, a head on both coins, we have

$$\Pr(X \cup Y) = \frac{1}{2} + \frac{1}{2} - \frac{1}{4} = \frac{3}{4}.$$

As we have seen, $X \cup Y$ can also be expressed as "at least one head." We have labeled this event W and found its probability to be 3/4 using the sample point techniques.

Frequently we may be interested in the combination of more than two events. As an illustration, suppose a passenger is waiting at a bus stop at which seven different buses arrive, on the average with equal frequency. Of these seven buses, four will take him to his home. What is the probability that the first bus which arrives will be one of those needed by the passenger?

Label the buses b_i and the event of the arrival of this bus by B_i for i = 1, 2, ..., 7. Suppose buses b_1, b_4, b_5, b_7 will take the passenger to his home. Then

$$\Pr\{\text{first bus to arrive is one needed by the passenger}\}$$
$$= \Pr(B_1 \cup B_4 \cup B_5 \cup B_7).$$

Since the events $B_1, B_2, \ldots, B_7$ are m.e. events, the determination of this probability is an extension of Axiom 3.

Addition Law for k Mutually Exclusive Events: *If* $A_1, A_2, \ldots, A_k$ *are* k *mutually exclusive events in* S, *then the probability of their union is*

$$\Pr(A_1 \cup A_2 \cup \cdots \cup A_k) = \sum_{i=1}^{k} \Pr(A_i).$$

Proof:

(i) This is clearly an extension of Axiom 3. Examination of a Venn diagram for k m.e. events shows that the probability of the union is $\sum_{i=1}^{k} \Pr(A_i) = \Pr(A_1) + \Pr(A_2) + \cdots + \Pr(A_k)$ since the events have no points in common.*

(ii) In this case the formal proof is based on mathematical induction.†

---

*The student should see Appendix A for a discussion of summation notation.

†The basic ideas of proof by mathematical induction are contained in Appendix C.

Verification:

For k = 1 we have the identity $Pr(A_1) = Pr(A_1)$. When k = 2, $Pr(A_1 \cup A_2) = Pr(A_1) + Pr(A_2)$ by Axiom 3. For k = 3, we wish to show $Pr(A_1 \cup A_2 \cup A_3) = Pr(A_1) + Pr(A_2) + Pr(A_3)$. Since we have an axiom concerning only two m.e. events, we need to write $A_1 \cup A_2 \cup A_3$ in terms of two m.e. events. Let $A_1 \cup A_2 = C$, then $A_1 \cup A_2 \cup A_3 = C \cup A_3$. Applying Axiom 3, we have $Pr(C \cup A_3) = Pr(C) + Pr(A_3)$. Resubstituting for C and again using Axiom 3, we have $Pr(A_1 \cup A_2 \cup A_3) = Pr(A_1) + Pr(A_2) + Pr(A_3)$, which is what we wished to show.

Extension:

We now assume the law to be true for k = n and based on this assumption we will prove it is for k = n + 1. the line of proof is analogous to that given for k = 3. Let $B = A_1 \cup A_2 \cup \cdots \cup A_n$. Then since B and A_{n+1} are m.e., we can use Axiom 3 to obtain $Pr(B \cup A_{n+1}) = Pr(B) + Pr(A_{n+1})$. By assumption, however, $Pr(B) = \sum_{i=1}^{n} Pr(A_i)$. Thus $Pr(A_1 \cup A_2 \cup \cdots \cup A_{n+1}) = \sum_{k=1}^{n+1} Pr(A_i)$.

Conclusion:

This law is true for all positive integral values of k. □

Returning to the passenger waiting for the bus, we see that

$$Pr(B_1 \cup B_4 \cup B_5 \cup B_7) = Pr(B_1) + Pr(B_4) + Pr(B_5) + Pr(B_7).$$

Since in this case the events are equally likely, the probability that at least one occurs is

$$Pr(B_1 \cup B_4 \cup B_5 \cup B_7) = \frac{4}{7}.$$

If $A_1, A_2, \ldots, A_k$ had formed a partition of S, what is the probability of their union? Here we have the special case in which $(A_1 \cup A_2 \cup \cdots \cup A_k) = S$ and $Pr(S) = 1$ by Axiom 2. Verbally this means that one of these k events must occur. In the penny-nickel example, the events U, V, Z form a partition of S. It is quite clear that the probability of their union is 1, since one of these events must occur. If any one of the buses could have taken the passenger to his home, then

$$Pr(B_1 \cup B_2 \cup \cdots \cup B_7) = Pr(S)$$

which is clearly 1.

Now consider the union of more than two events which are not m.e. How might we extend the addition law? If we are interested, in particular in the union of three events, we may find that it is useful to consider a Venn diagram depicting mutually exclusive regions.

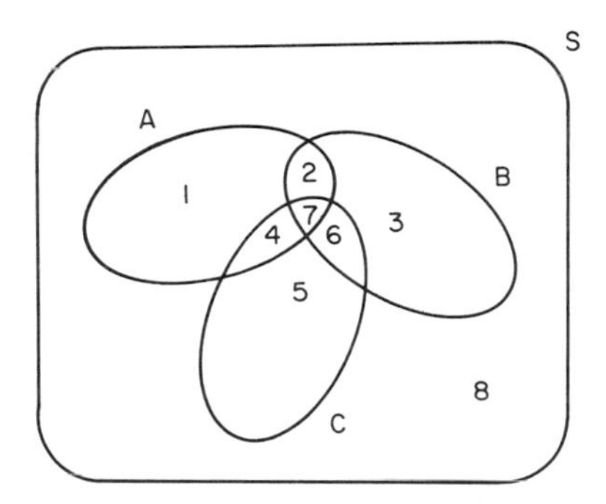

In this diagram the circles represent the three events A, B, C and the numbered regions correspond to the following m.e. events:

Region	*Event*
1	$A \cap \bar{B} \cap \bar{C}$
2	$A \cap B \cap \bar{C}$
3	$\bar{A} \cap B \cap \bar{C}$
4	$A \cap \bar{B} \cap C$
5	$\bar{A} \cap \bar{B} \cap C$
6	$\bar{A} \cap B \cap C$
7	$A \cap B \cap C$
8	$\bar{A} \cap \bar{B} \cap \bar{C}$

Since $A \cup B \cup C$ is the union of the first seven mutually exclusive events, we can determine the probability of the union of A, B and C by summing the probabilities of these seven events. In most cases, however, these probabilities will not be known.

Recall that when we talk about the union of events, we actually are interested in the probability that at least one event occurs. In keeping with this idea, the event $A \cup B \cup C$ will occur if one of the individual events A, B, C occurs or if any pair of the events occurs or if all three of the events occur simultaneously. We wish to develop a rule which depends only on the probabilities associated with the events: A, B, C, $A \cap B$, $A \cap C$, $B \cap C$, $A \cap B \cap C$.

> Addition Law for Three Events: *If* A, B *and* C *are any three events in* S, *then the probability of their union is*
>
> $$\begin{aligned}\Pr(A \cup B \cup C) &= \Pr(A) + \Pr(B) + \Pr(C) - \Pr(A \cap B) \\ &\quad - \Pr(A \cap C) - \Pr(B \cap C) + \Pr(A \cap B \cap C).\end{aligned}$$

Proof:

Let the set $(A \cup B)$ be equal to D. Then applying the addition law for two events, we have

$$\Pr(D \cup C) = \Pr(D) + \Pr(C) - \Pr(D \cap C).$$

Resubstitution for D gives

$$\Pr(A \cup B \cup C) = \Pr(A \cup B) + \Pr(C) - \Pr[(A \cup B) \cap C].$$

Again using the addition law,

$$\begin{aligned}\Pr(A \cup B \cup C) &= \Pr(A) + \Pr(B) - \Pr(A \cap B) \\ &\quad + \Pr(C) - \Pr[(A \cup B) \cap C].\end{aligned}$$

Now consider the probability of the event $(A \cup B)$ intersecting with C. The distributive law of sets gives

$$(A \cup B) \cap C = (A \cap C) \cup (B \cap C).$$

These two sets are <u>not</u> disjoint; their intersection is given by $A \cap B \cap C$. Thus

$$\Pr[(A \cap C) \cup (B \cap C)] = \Pr(A \cap C) + \Pr(B \cap C) - \Pr(A \cap B \cap C).$$

Substitution leads to

$$\Pr(A \cup B \cup C) = \Pr(A) + \Pr(B) + \Pr(C) - \Pr(A \cap B) - \Pr(A \cap C) - \Pr(B \cap C) + \Pr(A \cap B \cap C). \quad \square$$

Frequently we are interested in determining the probability that an event does <u>not occur</u>.

> Probability of the Complement: *If $\bar{A}$ is the complement of an event A in S, then*
>
> $$\Pr(\bar{A}) = 1 - \Pr(A).$$

Proof:

We know that A and $\bar{A}$ form a partition of S; hence, $\Pr(A \cup \bar{A}) = \Pr(S) = 1$. Since A and $\bar{A}$ are m.e., we can write $\Pr(A \cup \bar{A}) = \Pr(\bar{A}) + \Pr(A)$ by Axiom 3. Thus solving for $\Pr(\bar{A})$, we have $\Pr(\bar{A}) = 1 - \Pr(A)$. $\square$

We have used Venn diagrams to depict the combination of events. We may also find it useful to relate two events A and B by using a device known as a two-way table:

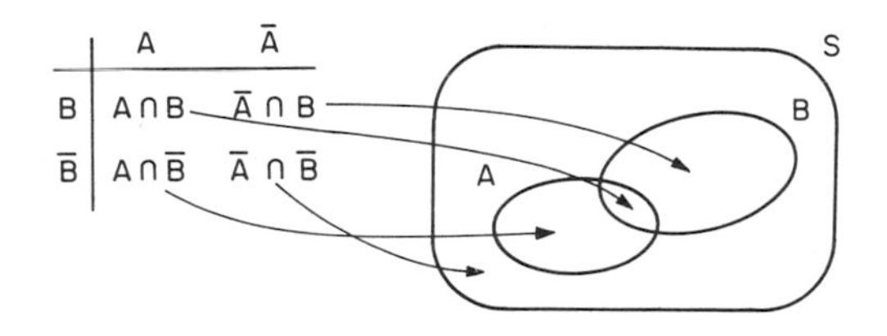

The entries within the two-way table depict the intersections of the events. The above diagram shows the relationship between the Venn diagram and the two-way table.

2-5 Finding Probabilities

The previous sections of this chapter contain definitions of basic terms useful in discussion of random experiments and elementary laws for determining probabilities of events. In this section several examples are given to illustrate the use of these definitions and laws.

Example 1*:* A commuter driving to work daily passes five traffic lights. For 100 days he observed the number of red lights which he encountered with the following relative frequencies:

Number of Red Lights	0	1	2	3	4	5
Relative Frequency	.05	.25	.37	.25	.05	.03

Find the probability of his encountering (a) at least three red lights, (b) more than three red lights, (c) at most 3 red lights.
Solution: We will assume that these relative frequencies can be used to approximate probabilities. Each of the events of interest in (a), (b) and (c) can be decomposed into the union of m.e. events. (a) The event "at least three red lights" is the union of three mutually exclusive events:

$$\Pr(\text{at least three red lights}) = \Pr(3) + \Pr(4) + \Pr(5) = .33.$$

(b) $$\Pr(\text{more then three red lights}) = \Pr(4) + \Pr(5) = .08.$$

(c) $$\Pr(\text{at most three lights}) = \Pr(0) + \Pr(1) + \Pr(2) + \Pr(3) = .92.$$

This event can also be determined as the complement of the event "more than three red lights"; hence,

$$\Pr(\text{at most three red lights}) = 1 - .08 = .92.$$

Example 2: Suppose a student is taking two mathematics courses (I, II) in summer school. Let A be the event that he passes course I and B be the event that he passes course II. He feels that Pr(A) = .8, Pr(B) = .9 and Pr(A ∩ B) = .75.

(a) Describe an appropriate sample space for this experiment.

(b) Using Venn diagrams, pictorially represent S.

(c) Describe in <u>words</u> the events: (i) $A \cup B$, (ii) $\bar{A} \cup \bar{B}$, (iii) $A \cap \bar{B}$, (iv) $\overline{(A \cup B)}$.

(d) Find the probabilities of the events in part (c).

Solution: We might use the ordered pair (x_1, x_2) to represent passing or failing courses I and II, respectively. Let $x_i = 1$ designate passing and $x_i = 0$ failing. Then S = {(1, 1), (1, 0), (0, 1), (0, 0)}. (b) A pictorial representation of S is given by:

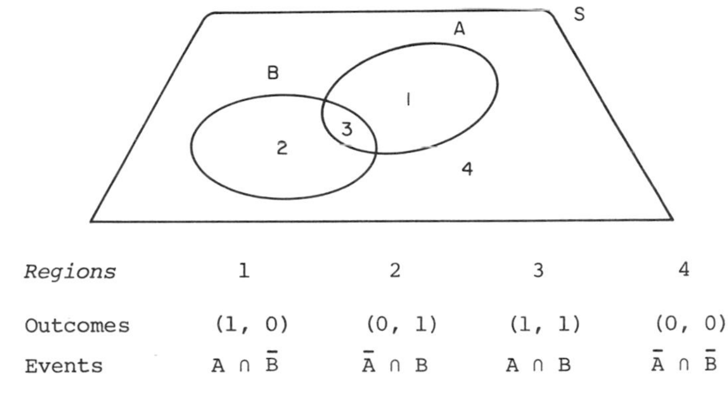

Regions	1	2	3	4
Outcomes	(1, 0)	(0, 1)	(1, 1)	(0, 0)
Events	$A \cap \bar{B}$	$\bar{A} \cap B$	$A \cap B$	$\bar{A} \cap \bar{B}$

(c)

(i) $A \cup B$: Passing at least one course; alternatively, passing course I or course II or both (includes regions 1, 2 and 3).

(ii) $\bar{A} \cup \bar{B}$: Failing at least one course; alternatively, failing course I or course II or both (includes regions 1, 2 and 4).

(iii) $A \cap \bar{B}$: Passing course I and failing course II (region I).

(iv) $\overline{A \cup B}$: Failing both courses; alternatively, passing neither course (region 4).

(d) The problem specified Pr(A), Pr(B) and $Pr(A \cap B)$. How can we determine the other intersection probabilities? We know that

$$Pr(B) = Pr(A \cap B) + Pr(\bar{A} \cap B)$$

which implies that

$$Pr(\bar{A} \cap B) = Pr(B) - Pr(A \cap B) = .9 - .75 = .15.$$

Similarly

$$Pr(A \cap \bar{B}) = Pr(A) - Pr(A \cap B) = .8 - .75 = .05$$
$$Pr(\bar{A} \cap \bar{B}) = Pr(\bar{A}) - Pr(\bar{A} \cap B) = .20 - .15 = .05.$$

It follows that

(i) $Pr(A \cup B) = Pr(A) + Pr(B) - Pr(A \cap B) = .95$ or
$Pr(A \cup B) = Pr(A \cap B) + Pr(A \cap \bar{B}) + Pr(\bar{A} \cap B) = .95.$

(ii) $Pr(\bar{A} \cup \bar{B}) = Pr(\bar{A}) + Pr(\bar{B}) - Pr(\bar{A} \cap \bar{B}) = .25$

(iii) $Pr(A \cap \bar{B}) = .05$. We found this by noting that $A = (A \cap \bar{B}) \cup (A \cap B)$ with $(A \cap \bar{B})$ and $(A \cap B)$ m.e. Since we know Pr(A) and $Pr(A \cap B)$, we can solve for $Pr(A \cap \bar{B})$.

(iv) $Pr(\overline{A \cup B}) = 1 - Pr(A \cup B) = .05.$

We may find it useful to use a two-way table to summarize the probabilities of the joint events:

	A	$\bar{A}$	
B	.75	.15	.90
$\bar{B}$	.05	.05	.10
	.80	.20	

The marginal probabilities indicate Pr(A), $Pr(\bar{A})$, Pr(B) and $Pr(\bar{B})$; those within the four cells are the intersection probabilities.

Example 3: On a surprise quiz with two questions, an instructor gave one multiple choice question with three alternatives a, b and c and one true-false type. Suppose a student is completely unprepared for the quiz, so he decides to answer at random.

(a) Describe a suitable sample space.

(b) If the correct answer is (b, T), find the probability that

(i) Both questions are answered correctly,

(ii) The first question is answered correctly,

(iii) At least one is answered ocrrectly,

(iv) At most one is answered correctly.

Solution: (a) A possible sample space consists of the six equally likely ordered pairs: (a, T), (a, F), (b, T), (b, F), (c, T), (c, F). (b) We can find these probabilities using the sample-point approach or the probability laws.

(i) Pr(correct on both) = Prob(b, T) = 1/6.

(ii) Pr(first question is correct) = Pr{(b, T) or (b, F)} = 1/3.

(iii) Pr(at least one is correct) = Pr{(b, T) or (b, F) or (a, T) or (c, T)} = 2/3. Alternatively, Pr(at least one is correct) = Pr(first is correct) + Pr(second is correct) - Pr(both are correct) = 1/3 + 1/2 - 1/6 = 2/3.

(iv) Pr(at most one is correct) = Pr{(a, T) or (a, F) or (b, F) or (c, T) or (c, F)} = 5/6. Alternatively, using the complement law, Pr(at most one is correct) = 1 - Pr(both correct) = 5/6.

Problems

1. A sample space consists of five simple events E_1, E_2, E_3, E_4 and E_5. Determine for each of the following whether the conditions of a proper probability model are satisfied:

   (a) $Pr(E_1) = .01$, $Pr(E_2) = .03$, $Pr(E_3) = .11$, $Pr(E_4) = .35$, $Pr(E_5) = .50$.

   (b) $Pr(E_1) = -.01$, $Pr(E_2) = .03$, $Pr(E_3) = .11$, $Pr(E_4) = .35$, $Pr(E_5) = .50$.

   (c) $Pr(E_1) = .80$, $Pr(E_2) = .00$, $Pr(E_3) = .10$, $Pr(E_4) = .05$, $Pr(E_5) = .005$.

   (d) $Pr(E_1) = .20$, $Pr(E_2) = .20$, $Pr(E_3) = .10$, $Pr(E_4) = .60$, $Pr(E_5) = .005$.

2. Let S consist of six simple events $E_1, E_2, \ldots, E_6$ with $Pr(E_1) = .10$, $Pr(E_2) = .25$, $Pr(E_3) = .15$, $Pr(E_4) = .20$, $Pr(E_5) = .10$, $Pr(E_6) = .20$.
   (a) If the events: $A = \{E_1, E_2, E_6\}$
   $B = \{E_3, E_4\}$
   $C = \{E_5, E_6\}$
   find the probabilities of the events $\bar{A}$, $\bar{C}$, $A \cup C$, $B \cup C$, and $A \cup B \cup C$.
   (b) Show that $Pr(A \cup B) = Pr(A) + Pr(B)$. Why is this true?
   (c) Does $Pr(A \cup C) = Pr(A) + Pr(C)$? Why or why not?

3. If S is a sample space containing n simple events and A is an event which contains m simple events, is it true that $Pr(A) = m/n$? Why or why not?

4. Let S contain six simple events $E_1, E_2, \ldots, E_6$ with $Pr(E_1) = Pr(E_2) = Pr(E_3) = Pr(E_4) = Pr(E_5) = Pr(E_6)$. Repeat parts (a), (b) and (c) in Problem 2.

5. A survey of families with three children is undertaken. Two possible sample spaces are suggested:
   S_1 = {BBB, BBG, BGB, GBB, BGG, GBG, GGB, GGG}
   S_2 = {Zero B, 1B, 2B, 3B}.

   Here S_1 preserves the birth order while S_2 records only the total number of boys.
   (a) Suppose we assume that each outcome in the sample space (for either S_1 or S_2) is equally likely. Show that the two models are not consistent in that the probabilities for certain events are different.
   (b) How might we demonstrate which probability model is more applicable?
   (c) Show that assigning probabilities to S_1 uniquely determine the probabilities in S_2, whereas given the probabilities in S_2 there are many possibilities for the probabilities in S_1.

6. In many dice games the player throws two dice and the total number of dots facing up is observed. Suppose one red die and one green die are thrown.

   (a) Describe an appropriate sample space for this experiment using: (i) A tabular form
   (ii) Set notation

   (b) Consider the events:

   A: Throwing a double

   B: Score on green die at least two more than that on red

   C: Sum equal to eight

   Describe each event as a subset of S.

   (c) If r is the outcome on the red die and g is the outcome on the green describe in <u>words</u> the following algebraically described events:

   (i) $r = 2g$

   (ii) $r + g > 5$

   (iii) $r \neq g$

   (d) If each of the outcomes in S is equally likely, find the probability of the events described in parts (b) and (c).

7. Referring again to the two-dice problem discussed in Problem 6, show that:

   $\Pr\{|r - g| = 1 \text{ or } 2\} = \Pr\{|r - g| = 0, 3, 4 \text{ or } 5\}.$

8. (a) Prove that if A is a subset of B, then $\Pr(A) \leq \Pr(B)$.

   (b) In this case describe in <u>words</u> the event (B - A). Give a realistic example in which we might be interested in the event (B - A).

9. Prove that if A is a subset of B, then

$$\Pr(B - A) = \Pr(B) - \Pr(A).$$

10. Using the result in Problem 9, provide an alternative proof for the addition law for two events.

11. (a) Show that $\Pr(E \cup F \cup G) = \Pr(E) + \Pr(F \cap \bar{E}) + \Pr(\bar{E} \cap \bar{F} \cap G)$.

    (b) Using the results of Problem 9 and part (a), prove that $\Pr(E \cup F \cup G) = \Pr(E) + \Pr(F) + \Pr(G) - \Pr(E \cap F) - \Pr(E \cap G) - \Pr(F \cap G) + \Pr(E \cap F \cap G)$.

12. Show that the probability of occurrence of at least one of the several events never exceeds the sum of the probabilities of these events; for example

$$\Pr(A \cup B) \leq \Pr(A) + \Pr(B).$$

When does the equality hold?

13. An airport limousine service has two cars. The larger car can carry five passengers and the smaller three. Suppose we are interested in the number of passengers each is carrying.

    (a) Enumerate the points in a sample space. [Hint: It may be useful to represent a typical point in S by (x, y) where x is the number in the larger car and y is the number in the smaller car.]

    (b) Consider the following events:

    D: At least one car is empty

    E: Together the cars carry two, four or six passengers

    F: The cars carry the same number of passengers

    Express the following events as subsets of S.

    (i) D, (ii) E, (iii) F, (iv) $E \cap F$, (v) $E \cup F$, (vi) $D \cap F$, (vii) $D \cup F$.

    (c) Describe the events in (iv) through (vii) of (b) in words.

    (d) Assuming each point in S is equally likely, find the probability associated with the events (i) through (vii) in (b).

    (e) Find the probabilities for the events (v) and (vii) using the addition law.

14. For the two-dice experiment described in Problem 6, let A be the event that $r \leq 2$ and B the event that $g \geq 4$.
    (a) Find the probabilities of the events $\bar{A}$, A, $\bar{B}$, B, $A \cap B$, $\bar{A} \cup \bar{B}$, $A \cup B$, $\bar{A} \cap \bar{B}$.
    (b) Describe in <u>words</u> the events $A \cap B$, $A \cup B$, $\bar{A} \cup \bar{B}$ and $\bar{A} \cap \bar{B}$.

15. (a) Using Venn diagrams illustrate DeMorgan's laws that for any sets E and F
        (i) $\overline{(E \cup F)} = \bar{E} \cap \bar{F}$
        (ii) $\overline{(E \cap F)} = \bar{E} \cup \bar{F}$
    (b) Recall the events A and B described in Problem 14. Using the results of part (a), discuss the relationship between the events $A \cap B$, $A \cup B$, $\bar{A} \cup \bar{B}$, $\bar{A} \cap \bar{B}$ in <u>words</u> and in probabilities.

16. Generalize DeMorgan's laws to three sets:
    (a) $\overline{(E \cup F \cup G)} = \bar{E} \cap \bar{F} \cap \bar{G}$
    (b) $\overline{(E \cap F \cap G)} = \bar{E} \cup \bar{F} \cup \bar{G}$.

17. A sample survey was undertaken to investigate which papers (A, B, C) people read. In a sample of 100 people the following results were obtained:

60 read A	32 read A and B	30 read A, B and C
40 read B	45 read A and C	
70 read C	38 read B and C	

If a person is selected at random from this sample, determine the probability that:

(a) He reads only newspaper A.
(b) He reads only newspaper B.
(c) He reads only newspaper C.
(d) He reads at least two newspapers.
(e) He reads at most one newspaper.
(f) He does not read any newspaper.

18. A committee of three is selected from six people: A, B, C, D, E and F.
    (a) Specify a suitable sample space S and make an assignment of probabilities to the simple events in S.
    (b) Find the probability that:
        (i) A is selected.
        (ii) B is selected.
        (iii) A and B are selected.
        (iv) A or B is selected.
        (v) A is not selected
        (vi) Neither A nor B is selected.
    (c) How would the answers in part (b) have changed if we had considered persons B and C rather than A and B?

19. A jar contains an equal number of red, white and pink chips which are indistinguishable except for color. A chip is drawn and its color is noted. This chip is then replaced and a second chip is drawn.
    (a) Using R, W and P to represent the colors, discuss a possible sample space for this experiment.
    (b) Consider the following events:
        A: At least one red
        B: One pink and one red
        C: At most one white
        D: The second draw is a white one
        E: No white at all
        List the points in S which belong to the events A, B, C, D, E.
    (c) List the points in the events $A \cap C$, $A \cup C$, $B \cap E$, $B \cup D$, $\bar{E}$.
    (d) Describe the events in part (c) in <u>words</u>.
    (e) Why is it reasonable to assign an equal probability to each point in S?
    (f) Assuming each point in S is equally likely, find the probability of the events in (b) and (c).

20. An assembly line operation has two shifts: night (N) and day ($\bar{N}$). Articles coming off the assembly line may or may not be defective. Suppose that the probability that an item was produced on the night shift is .40 and that the probability that it is defective (D) is .01. We further know that the probability that the item is both defective and produced on the night shift is .002.
    (a) Describe an appropriate sample space using a two-way table.
    (b) Find $\Pr(\bar{N} \cap \bar{D})$, $\Pr(N \cap \bar{D})$, $\Pr(\bar{N} \cap D)$. Show all your work.

21. In the next Great Turtle Race, there are four entries. To introduce a little excitement into the affair, the four owners A, B, C, D decide to place their bets randomly in the following manner. The names of the four turtles are placed in a hat and each owner selects a name at random without replacement.
    (a) Set up an appropriate sample space to describe the assignment of turtles to bettors and assign appropriate probabilties to the simple events.
    (b) Express the following events as subsets of the sample space and calculate their probability of occurrence.
        (i) All owners get their own turtles.
        (ii) No owner gets his own turtle.
        (iii) The unfortunate A chooses Pokey, who always loses.
        (iv) A get B's turtle and B gets A's.

22. Two men and three women are in a chess tournament. Those of the same sex have an equal opportunity of winning but each man is twice as likely to win as any woman.
    (a) Describe an appropriate sample space for this experiment (winning the tournament).
    (b) What is the probability associated with each point in S?

(c) Find the probability that a woman wins the tournament.

(d) If two of the five are a married couple, find the probability that the tournament is won by one of them.

23. Consider again the sample space S_1 described in Problem 5.

(a) Designate the following events symbolically as subsets of S_1 and assuming each outcome is equally likely find their probabilities:

(i) Exactly one boy

(ii) At least one boy

(iii) At most one boy

(iv) More boys than girls

(v) The oldest is a boy

(vi) More girls than boys.

(b) Describe in words the complements of the events (i), (ii) and (iii). Find the probabilities of these events.

(c) Using the events of part (a), describe a partition of S_1.

(d) How would you describe in words the intersection of the following pairs of events?

(i) and (ii)

(ii) and (iv)

(v) and (vi)

Find the probabilities of these intersections.

(e) Find a pair of mutually exclusive events from the events listed in part (a).

24. A high school senior applies for admission to college A and to college B. He estimates that the probability of his being admitted to A is .7, that his application will be rejected at B with probability .5, and that the probability of at least one of his applications being rejected is .6. What is the probability that he will be admitted to at least one of the colleges?

25. A school division is interested in examining a possible relationship between nursery school attendance and reading readiness. At the beginning of the kindergarten year, they found that 90% of the students had attended nursery school, 60% passed the reading readiness test and 55% had both attended nursery school and passed the reading readiness test. Let R be the event that a student chosen at random has passed the reading readiness test and N the event that he has attended nursery school.

    (a) Display these results in a two-way table.

    (b) Explain in <u>words</u> the events $(R \cap N)$, $(R \cap \bar{N})$, $(\bar{R} \cap \bar{N})$ and $(\bar{R} \cap N)$.

    (c) Explain why $Pr(R) = Pr(R \cap N) + Pr(R \cap \bar{N})$.

    (d) Show that $Pr(R \cap N) = 1 - Pr(R \cup \bar{N})$. Interpret this result.

26. Suppose that in a class of 35 students the instructor daily chooses one student and asks him a question.

    (a) Let C_i be the event that the ith $(i = 1, 2, \ldots, 35)$ student is called upon to answer. Are the C_i's mutually exclusive? Do the C_i's form a partition of the sample space?

    (b) If Q is the event that the question is answered correctly, how might you depict the event Q? What probabilities would you need to know in order to determine $Pr(Q)$?

27. A pizza maker may or may not include onion, green pepper or pepperoni on his pizzas.

    (a) Enumerate the eight possible types of pizza he may make. Label the outcomes $E_1, E_2, \ldots, E_8$.

    (b) Let: A_1 be the event that the pizza has onion
    A_2 be the event that the pizza has green pepper
    A_3 be the event that the pizza has pepperoni

    Show that if we know the probabilities of the events A_1,

(b) A_2, A_3, $(A_1 \cup A_2)$, $(A_1 \cup A_3)$, $(A_2 \cup A_3)$ and $(\bar{A}_1 \cap \bar{A}_2 \cap \bar{A}_3)$, we can determine the probabilities of $E_1, E_2, \ldots, E_8$.

(c) Show that a knowledge of $\Pr(E_i)$ for $i = 1, 2, \ldots, 8$ allows us to calculate the probability of any event.

(d) If we know $\Pr(A_1) = 1/2$, $\Pr(A_2) = 1/2$, $\Pr(A_3) = 1/2$, $\Pr(A_1 \cup A_2) = 3/4$, $\Pr(A_1 \cup A_3) = 3/4$, $\Pr(A_2 \cup A_3) = 3/4$ and $\Pr(\bar{A}_1 \cap \bar{A}_2 \cap \bar{A}_3) = 1/8$, determine $\Pr(E_i)$ for $i = 1, 2, \ldots, 8$ using the formulas developed in part (b).

Chapter 3

Counting Procedures and Their Applications in Computing Probabilities

3-1 The Need for Counting Techniques: The Uniform Model

Often it may be reasonable to assign equal probability to each simple event in the sample space. If we adopt this assignment of probabilities, we say that the *probability model* is *uniform*.

In this situation the calculation of probabilities of particular events is simplified. If the sample space S contains k simple events, then using a uniform probability model we assign probability 1/k to each point in S. To determine the probability of an event E we need

1. the number of possible outcomes in S,
2. the number of outcomes corresponding to E.

In this special case Definition 2·4 becomes

$$\Pr(E) = \frac{\text{number of outcomes corresponding to E}}{\text{number of possible outcomes in S}} .$$

Frequently it may be possible to enumerate fully all the sample points in S and then count how many of them correspond to the event E.

Suppose a class consists of just three students. The instructor always calls on each student once and only once during each class. Let us label the students 1, 2 and 3. Then we can easily enumerate the points in S as

$$S = \{(1, 2, 3), (1, 3, 2), (2, 1, 3), (2, 3, 1), (3, 1, 2), (3, 2, 1)\},$$

where the point (x, y, z) is an ordered triplet designating person x as the first respondent, person y as the second respondent and person z as the third respondent. We can also illustrate the sample space pictorially using a tree diagram:

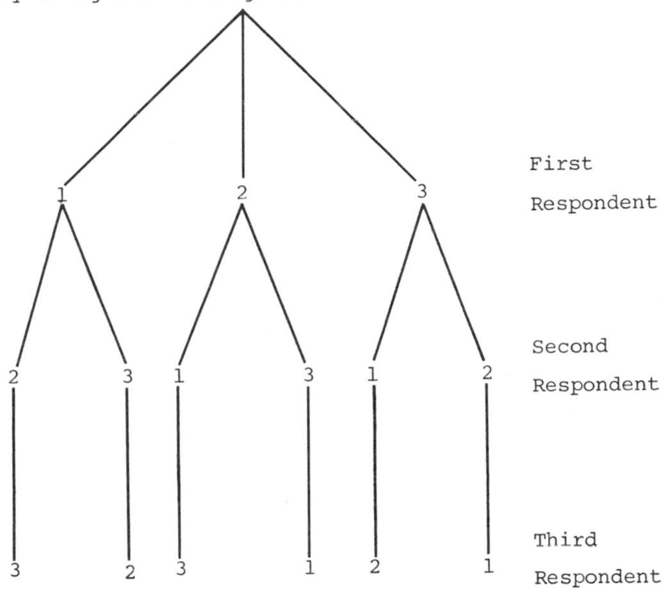

A sample point is then found by reading along a path of branches. Note that in using the tree diagram each sequence of branches represents a different arrangement or order.

Assuming that the instructor chooses the students at random (no preference for a particular student), it would seem reasonable to adopt a uniform probability model and assign probability 1/6 to each point in S. Now let A, the event of interest, be

A = event that John is selected last.

Immediately you will ask whether John is student 1, 2 or 3. An examination of the enumerated sample space (or the tree diagram) shows that regardless of his number, John has the same probability of being selected last; namely,

$$\Pr(A) = \frac{2}{6} = \frac{1}{3} .$$

We shall see that this is always true if the selection is at random.

It would be most unusual for a class to consist of only three students. We shall see that total enumeration of the sample space becomes more complicated even if we increase the class size to five students. To deal with these situations in which the sample space contains a large number of points, we will need to have an understanding of basic counting or combinatorial procedures.

3-2 Counting Procedures Involving Order Restrictions

Extending our example, suppose the instructor calls on exactly three students out of a class of five. In order to apply the uniform probability model, we need to know how many points there are in S. Again each point in S is an ordered triplet. Now, however, the first question can be answered by one of five students, the second question (after the first has been answered) by one of four students and the third question (after the first and second are answered) by one of three students. Thus S contains

$$5 \times 4 \times 3 = 60 \text{ points.}$$

If we view this as a tree diagram, there would be five main branches, four secondary from each main one and three tertiary from each secondary one. In all there would be 60 sequences of branches corresponding to the possible outcomes in S.

We will now generalize these results and state them formally as the Multiplication Principle or Fundamental Rule of Counting.

Multiplication Principle: *If an operation consists of a sequence of* k *separate steps of which the first can be performed in* n_1 *ways, followed by the second in* n_2 *ways, and so on until the* kth *can be performed in* n_k *ways, then the operation consisting of* k *steps can be performed by*

$$n_1 \times n_2 \times \cdots \times n_k$$

ways.

In our example the points in S were represented by ordered triplets; that is, the point (3, 5, 1) is different from (5, 3, 1). The same three people were called upon to respond, but the order of response is different. We shall refer to such an ordering or arrangement as a permutation.

Definition 3·1: *A* permutation *of a number of objects is any arrangement of these objects in a definite order.*

When the class contained only three and each was called on exactly once, we found that there were 3 × 2 × 1 ways in which the students might be called upon. In general, if the class had consisted of n students and all of them had been called upon, then the responses could have taken place in

$$n \times (n - 1) \times (n - 2) \times \cdots \times 3 \times 2 \times 1$$

ways. We will find it useful to abbreviate this product by n!, which is read "n factorial."

Theorem 3·1: *The number of permutations of a set of* n *distinct objects, taken all together, is* n!.

Proof:

The operation here consists of filling n spaces. Applying the multiplication principle, we see that the first space can be filled in n ways; after filling the first, the second can be filled in (n - 1) ways and so on. Thus the n spaces can be filled in n! ways. □

By definition 0! is taken to be equal to 1.

In the second part of our example only three people were to be called on from a class of ten; that is, we are interested in an ordered subset.

> Definition 3·2: *An arrangement of* r *distinct objects taken from a set of* n *distinct objects is called a* permutation of n objects taken r at a time. *The total number of such orderings will be symbolized by* ${}_nP_r$.

We found in our example that ${}_5P_3$ was equal to $5 \times 4 \times 3 = 60$. We will now generalize this result to ${}_nP_r$.

> Theorem 3·2: *The number of permutations of* n *distinct objects taken* r *at a time, without repetitions, is*
>
> $$n!/(n - r)!$$

Proof:

Applying the multiplication principle, we see that the first position can be filled in n ways, the second position (after the first is filled) in (n - 1) ways, and so forth to the rth position which can be filled in n - (r - 1) ways. Thus the r positions can be filled in

$${}_nP_r = n(n - 1)(n - 2) \cdots (n - r + 1)$$

ways. To express the right-hand side (r.h.s.) in terms of factorials, we multiply by $(n - r)!/(n - r)!$ and obtain

$$
\begin{aligned}
{}_nP_r &= n(n-1) \cdots (n-r+1) \frac{(n-r)!}{(n-r)!} \\
&= \frac{n!}{(n-r)!}
\end{aligned}
$$

□

Note that ${}_nP_n$ is $n!$ which is the result given in Theorem 3·1.

Thus far we have always been discussing ordering of different or distinct items. Suppose that in the class of three students, two of them are identical twins. Clearly, since the instructor cannot distinguish between the twins, there will be fewer possible outcomes in S. The six points in S will be collapsed to three points.

$$
\left.\begin{array}{l}(1, 2, 3)\\(2, 1, 3)\end{array}\right\} \quad (T, T, 3)
$$

$$
\left.\begin{array}{l}(1, 3, 2)\\(2, 3, 1)\end{array}\right\} \quad (T, 3, T)
$$

$$
\left.\begin{array}{l}(3, 1, 2)\\(3, 2, 1)\end{array}\right\} \quad (3, T, T)
$$

If the T's were distinguishable, each of the permutations on the right-hand side would have consisted of the 2! orderings on the left. Let a be the number of permutations when not all elements are distinct. Then we can express the total number of permutations if all elements are distinct as

$$3! = a(2!)$$

Solving for a, we get

$$a = \frac{3!}{2!} = 3.$$

If we had n items with r objects alike, then the number of distinct permutations taking all n at a time is $n!/r!$.

Now consider a group of n people of which r are females and n - r are males. In how many ways can these people line up for theater tickets if we are interested only in distinguishing between sexes?

Let a be the number of such permutations when we distinguish only by sex. Then using the multiplication principle, we can express the total number of permutations if all elements are distinct as

$$n! = a[r!][(n - r)!]$$

Solving for a, we get

$$a = \frac{n!}{r!(n - r)!} .$$

Repeated application of this argument leads to the following theorem.

> Theorem 3·3: *In a set of* n *elements having* r_1 *elements of one type,* r_2 *of a second type and so on to* r_k *of a* kth *type, the number of distinct permutations of the* n *elements, taken all together, is*
>
> $$\frac{n!}{r_1!r_2! \cdots r_k!} ,$$
>
> *where* $\sum_{i=1}^{k} r_i = n$.

3-3 Counting Procedures Not Involving Order Restrictions

In our previous discussions we have been concerned with the order in which the students were called upon to answer. If we relax this condition of order, then how many different groups or subsets of three members can be chosen from the five? Recall that we described a typical point in the sample space as the ordered triplet (x, y, z). If we are interested only in the membership and do not consider order, then the six triplets

(x, y, z), (x, z, y), (y, x, z), (y, z, x), (z, x, y), (z, y, x) all have the same membership x, y, z. We shall use the notation (x y z) to designate an unordered point. An examination of S shows that there are 10 groups of six ordered triplets having the same membership. Thus there are 10 unordered triplets.

Definition 3·3: *A subset of* r *objects selected without regard to order from a set of* n *different objects is called a* combination *of* n *objects taken* r *at a time. The total number of such combinations will be designated by*

$$ {}_nC_r = \binom{n}{r} \quad \textit{for } r \leq n. $$

From our discussion we know that we should be able to evaluate the symbol $\binom{n}{r}$ from our knowledge of the number of ordered subsets.

Theorem 3·4: *The number of combinations of a set of* n *different objects taken* r *at a time is*

$$ {}_nC_r = \binom{n}{r} = \frac{n!}{r!(n-r)!} . $$

Proof:

The selection of r objects from a total of n is equivalent to partitioning the n objects into two groups: the r selected and the (n - r) remaining, with no regard to order within the groups. Let ${}_nC_r$ be the number of ways of partitioning the n items into two groups, ignoring order within the two groups. We know that the total number of orderings can also be expressed as

(number of ways of partitioning into two groups) ×
(number of ways of ordering each group).

Thus

$$ {}_nP_n = n! = {}_nC_r[r!][(n-r)!]. $$

Solving for ${}_nC_r$, we get

$${}_nC_r = \frac{n!}{r!(n - r)!} . \qquad \square$$

This result can easily be extended to partitioning of n items into k subsets.

> Theorem 3·5: *The number of ways of partitioning* n *distinct items into* k *groups of* n_1, n_2, ..., n_k, *respectively, is*
>
> $$N = \binom{n}{n_1 n_2 \cdots n_k} = \frac{n!}{n_1!n_2! \cdots n_k!}$$
>
> *where* $\sum_{i=1}^{k} n_i = n$.

Proof:

Using an argument analogous to that of Theorem 3·4, we have

$${}_nP_n = n! = N(n_1!n_2! \cdots n_k!).$$

Solving for N, we have

$$N = \frac{n!}{n_1!n_2! \cdots n_k!} . \qquad \square$$

Since we are not distinguishing among the possible orders in the subsets, this partitioning is equivalent to the number of permutations of n objects in which n_1 are of one type, n_2 of a second type and so on (see Theorem 3·3).

3-4 Applications of Counting Procedures

In this section we will discuss several examples which illustrate the counting techniques developed in the previous section of this chapter and the laws of probability developed in Chapter 2.

Example 1: In a bakery, cookies are decorated with at least one topping consisting of icing, sugar or candies. In how many ways can a cookie be decorated?

Solution: A baker can choose (C) or not choose ($\bar{C}$) each of the three toppings. Thus a cookie can be decorated in $2 \times 2 \times 2 = 8$ ways, but this includes the omitting of all three toppings. Since the cookie must have at least one topping, it can be decorated in $2 \times 2 \times 2 - 1 = 7$ ways.

Another way of attacking this problem is to consider that we have three toppings from which to choose. We want to find all the ways in which we can select one topping <u>or</u> two toppings <u>or</u> three toppings:

$$\binom{3}{1} + \binom{3}{2} + \binom{3}{3} = 3 + 3 + 1 = 7$$

Note that this is essentially finding all the possible subsets, excluding the null set, of a universal set consisting of three elements.

Example 2: A book shelf has seven books. Three of the books have identical red covers and four have identical white covers. In how many ways can at least one book be selected from the shelf?

Solution: Since the books with red covers are indistinguishable among themselves, their selection can be identified only by the <u>number of books</u> chosen: 0, 1, 2 or 3. That is, there are four ways to choose the red books. Similarly, the books with white covers can be selected in five ways. The books can be selected together in $4 \times 5 = 20$ ways. But this contains the selection of no red covered book and no white covered book. Thus at least one book can be selected in $4 \times 5 - 1 = 19$ ways.

Example 3: If the books were all different, in how many ways could a selection of at least one book be made?

Solution: Now this problem is analogous to Example 1. Here each book can be chosen or rejected; hence, there are $2^7 - 1 = 127$ selections containing at least one book.

Example 4: Suppose six people, labeled A through F, line up at random at a theater box office. (a) What is the probability that persons A and B are standing next to one another? (b) What is the probability that person E is not first in line?

Solution: A possible sample space for this experiment consists of all possible permutations of the letters A through F; that is, S consists of $6! = 720$ points. If we assume that the individuals line up at random, we can adopt the uniform model and assign each point in S a probability of 1/720. (a) Let V be the event that A and B are adjacent. Now this can occur in two mutually exclusive orderings: A followed by B or B followed by A. If we label these two events V_1 and V_2, respectively, then $V = V_1 \cup V_2$ and $\Pr(V) = \Pr(V_1) + \Pr(V_2)$. How many points in S correspond to the event V_1? If A and B must be together in this order, then we can treat them as being one item. Hence the number of individuals is effectively reduced to five and there are $5! = 120$ possible arrangements. Thus,

$$\Pr(V_1) = \frac{5!}{6!} = \frac{1}{6} .$$

By symmetry

$$\Pr(V_2) = \frac{1}{6} ;$$

hence,

$$\Pr(V) = \frac{1}{6} + \frac{1}{6} = \frac{1}{3} .$$

(b) This probability can be obtained most easily by considering the complement law. Let W be the event that E does not come first. Then the complement event $\bar{W}$ is E appearing first. If E's position is fixed in the first place, there are five positions to be filled. Thus

$$Pr(\bar{W}) = \frac{5!}{6!} = \frac{1}{6}$$

and

$$Pr(W) = 1 - Pr(\bar{W}) = \frac{5}{6}.$$

Example 5: A committee of nine members decides to select its officers by lot. From the nine names, three will be drawn successively. The first one drawn will be the chairperson, the second will be the vice-chairperson and the third the secretary. (a) How many possible outcomes are there for this selection of officers? (b) What probabilities would you assign to each of these outcomes? (c) If there is one woman on the committee, what is her chance of becoming the vice-chairperson? (d) What is the probability that the three oldest persons on the committee will be chosen as chairperson, vice-chairperson and secretary, respectively? (e) What is the probability that the three oldest persons will be chosen as officers but not necessarily in ranked order?

Solution: (a) Since we are interested in the order in which the selection is made, there will be ${}_9P_3 = 504$ outcomes in S. (b) A selection by lot implies that each outcome is equally likely; hence, we would assign a probability of 1/504 to each point in S. (c) A typical sample point belonging to this event is (x_1, W, x_3), where the W indicates that the second position is filled by the woman. In how many ways can the positions x_1 and x_3 be filled? At first glance you might think that we need only choose a subset of two people from the remaining eight; that is, ${}_8C_2$. Why is this not correct? Since we have considered order in finding the total number of points in S, we must also consider order in calculating the number of sample points belonging to the event of interest. The correct answer is ${}_8P_2$ and the probability is

$$\frac{{}_8P_2}{{}_9P_3} = \frac{1}{9}.$$

(d) In this case we wish to select the three oldest people in a particular order corresponding to the rank of their ages; hence, only one point in S corresponds to the event. The probability is 1/504. (e) This problem is similar to (d); however, now we are not insisting that the oldest person become chairperson, but only that he be included in the committee. Thus, there will be 3! orderings of committees made up of the three oldest members. Here the probability is

$$\frac{3!}{{}_9P_3} = \frac{1}{84}$$

Example 6: By mistake four "burned out" (B) light bulbs have been mixed up with 16 good (G) ones. If a random sample of three is drawn, what is the probability (a) all three will be good? (b) at least one is "burned out"? (c) at most two are "burned out"?
Solution: Here we are not concerned with order. We wish to take at random a subset of three from the 20; hence S contains ${}_{20}C_3 = 1140$ equally likely points. (a) Here we are restricting ourselves to all G bulbs. There are ${}_{16}C_3$ possible samples with all G bulbs. Thus

$$\Pr(\text{no B bulbs}) = \frac{{}_{16}C_3}{{}_{20}C_3} = \frac{28}{57}\ .$$

(b) "At least one B" is the complement of "no B." Hence

$$\Pr(\text{at least one B}) = \left(1 - \frac{{}_{16}C_3}{{}_{20}C_3}\right) = \frac{29}{57}\ .$$

(c) "At most two B's" is the union of the events "no B's," "one B", "two B's." Note that this is also the complement of "three B's."

$$\Pr(\text{three B's}) = \frac{{}_4C_3}{{}_{20}C_3} = \frac{1}{285}\ .$$

Therefore

$$\Pr(\text{at most two B's}) = 1 - \left(\frac{{}_4C_3}{{}_{20}C_3}\right) = \frac{284}{285}\ .$$

Example 7: Suppose a child is asked to rank three toys labeled A, B and C in order of preference. If he, like many small children, has difficulty making decisions and merely ranks the toys at random, each order can be assumed equally likely. What is the probability that (a) toy A is given first preference? (b) A is given second preference? (c) A is ranked first <u>or</u> second? (d) A is ranked first <u>and</u> B second?

Solution: (a) A possible sample space for this experiment is the 3! orderings of the letters A, B and C. Under the assumption of no preference, each of the six points in S is equally likely. If A is to be ranked first, then the first position is predetermined. The remaining two positions can be filled in 2! ways. Thus

$$\Pr(\text{A ranked first}) = \frac{2!}{3!} = \frac{1}{3}.$$

(b) An argument analogous to that in part (a) gives

$$\Pr(\text{A ranked second}) = \frac{1}{3}.$$

(c) Since the events "A ranked first" and "A ranked second" are mutually exclusive,

$$\Pr(\text{A ranked first } \underline{\text{or}} \text{ second}) = \frac{1}{3} + \frac{1}{3} = \frac{2}{3}.$$

(d) If A and B are to be fixed in this order in the first two positions, then the third position can be filled only by C and

$$\Pr(\text{A first } \underline{\text{and}} \text{ B second}) = \frac{1}{6}.$$

Example 8: A nutrition research project was undertaken to study the effect on flavor of defrosting (D) or not defrosting (N) meat prior to cooking. Each of four tasters was presented with two specimens of meat, one defrosted prior to cooking and the other not. The tasters were asked to choose the specimen which had the better flavor. (a) Discuss an appropriate sample space and assign probability to each point in S. (b) What is the probability that the first three tasters prefer "not defrosted?" (c) What is the probability that exactly three choose the "not defrosted" specimen? (c) What is the probability that at least three tasters choose "not defrosted?"

Solution: (a) At first glance we might think that we need only list the number of tasters, 0, 1, 2, 3 or 4, choosing D. If, in fact, these outcomes are chosen for describing S, care must be taken in assigning the probabilities. Even if we assume no preference for D or N, these outcomes are not equally likely. We also observe that in order to answer part (b), we need a more detailed sample space in which the outcome for each taster is identified. An appropriate description of an outcome would be the <u>ordered</u> four-tuple (x_1, x_2, x_3, x_4) where each x_i could be a D or N. Thus, S has $2^4 = 16$ equally likely outcomes. (b) We are looking for outcomes of the form (N, N, N, x_4). There will be two such outcomes:

$$\Pr(\text{first three tasters prefer N}) = \frac{2}{16} = \frac{1}{8}\,.$$

(c) The outcomes of interest here are the distinct permutations of 3 N's and a D: $4!/3! = 4$.

$$\Pr(\text{exactly 3 N's}) = \frac{4}{16} = \frac{1}{4}\,.$$

(d) The event "at least 3 N's" is the union of the mutually exclusive events: "exactly 3 N's" and "exactly 4 N's."

$$\Pr(\text{at least 3 N's}) = \frac{1}{16} + \frac{4}{16} = \frac{5}{16}\,.$$

Example 9: Joe and Bill each roll a die three times. Each of them notes the sum of the outcomes on his three rolls. Joe's score is 9 and Bill's is 10. Joe says that a 9 can be produced in six ways: 1 + 2 + 6, 1 + 3 + 5, 1 + 4 + 4, 2 + 2 + 5, 2 + 3 + 4, 3 + 3 + 3. Bill claims that a sum of 10 can also occur in six ways: 1 + 3 + 6, 1 + 4 + 5, 2 + 2 + 6, 2 + 3 + 5, 3 + 3 + 4, 4 + 4 + 2. Thus the boys conclude that the probabilities of getting a sum of 9 or a sum of 10 are equal. Do you agree?

Solution: There are six possible outcomes for each throw of the die; hence, S consists of $6 \times 6 \times 6 = 216$ ordered outcomes. If the die is balanced, we can assume each of these outcomes is equally likely. Using Joe's and Bill's argument we would conclude that

$$\Pr(\text{sum} = 9) = \Pr(\text{sum} = 10) = \frac{6}{216}.$$

If, however, we were to carry out this experiment, we would find that these probabilities are not supported by the observed relative frequencies. Why? Consider the sum of 9. The unordered outcomes (1 2 6), (1 3 5) and (2 3 4) can each occur in 3! ways. Since the outcomes (1 4 4) and (2 2 5) have only two distinguishable elements, each of them can occur in 3!/2! ways. As the 3's in the last outcome (3 3 3) are not distinguishable from one another, this outcome occurs in only one way. We then have

$$\Pr(\text{sum} = 9) = \frac{(3!)(3) + (3!/2!)(2) + 1}{216} = \frac{25}{216}.$$

A similar analysis for the sum = 10 gives

$$\Pr(\text{sum} = 10) = \frac{(3!)(3) + (3!/2!)(3)}{216} = \frac{27}{216}.$$

Bill's score is more probable!

This example illustrates the importance of distinguishing between ordered and unordered outcomes.

Example 10: A children's board game is laid out as a rectangular grid or lattice: six blocks long and four blocks wide. Pieces are moved either east or north from the starting point 0 to the finish

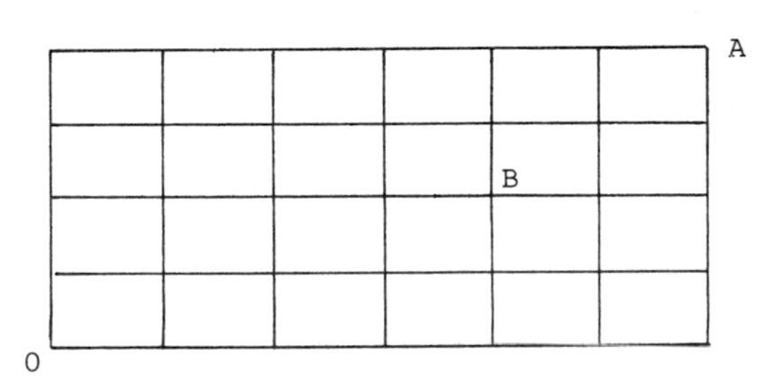

point A. (a) How many sequences of moves are possible? (b) If the player passes through point B, he gets a bonus. If each sequence is equally likely, what is the probability that a player passes through B on his way from 0 to A?

Solution: (a) Each path is some ordering of six moves east (E) and four moves north (N). Since the moves in either direction are not distinguishable within themselves, we are interested in the number of distinct permutations of six E's and four N's: $10!/6!4! = 210$.

(b) A player can move from 0 to B by moving east four times and north twice. Thus the number of paths from 0 to B is $6!/4!2! = 15$. Similarly the number of paths from B to A is $4!/2!2! = 6$. Then

$$\Pr(\text{B is included in path from 0 to A}) = \frac{(15)(6)}{210} = \frac{3}{7}.$$

Example 11: This example is a variation on the classic "birthday problem." In a class of 30 students, what is the probability that at least two of them have the same birthday (month and day)?

Solution: There are 365 (omitting leap years) possibilities for each person's birthday; hence, S contains 365^{30} points. A typical point in S is the ordered 30-tuple $(x_1, x_2, \ldots, x_{30})$. Assuming each point in S is equally likely, each point in S has probability $1/365^{30}$. Let A be the event that no two of the 30 people have the same birthday. The event of interest, "at least two people have the same birthday," is then the complement event $\bar{A}$. How many points in S lie in A? There will be ${}_{365}P_{30}$ points in A. Note that here again we must consider order since the total number of points in S were calculated under an order restriction; thus

$$\Pr(\bar{A}) = 1 - \Pr(A) = 1 - \frac{{}_{365}P_{30}}{365^{30}} = .706$$

How is this probability affected by changing the number of people in the room?

Number of People	5	10	20	23	30	40	60
$Pr(\bar{A})$	.027	.117	.411	.507	.706	.891	.994

What is the expression for this probability when there are k people in the room?

3-5 Occupancy Problems: Role of Distinguishability

In Examples 2 and 3 the basic concepts of "distinguishable" and "indistinguishable" elements were introduced. Here the covers of the seven books were of two colors; hence, we might represent the books symbolically as R_1, R_2, R_3, W_1, W_2, W_3, W_4. In this notation the seven books are clearly different and distinguishable from one another: R_1 is distinguishable from R_2 by its number, while R_1 is distinguishable from W_1 by its color. But if thsse same books are distinguishable only on the basis of their color, then the numerical subscripts disappear and the elements can be denoted symbolically as R, R, R, W, W, W, W.

What then is the role of distinguishability in constructing sample spaces? Should we always distinguish among elements whenever possible?

Suppose two patients A and B arrive at a Public Health Clinic. Each selects one of three doctors (I, II, III) at random. What is an appropriate sample space for this problem? What probability should be assigned to each point in the sample space?

It is clear that the patients A and B are distinguishable. Let the ordered pair (x, y) with x = I, II, III and y = I, II, III represent A's choice and B's choice, respectively. The sample space then consists of $3^2 = 9$ points: $S = \{(x, y) \mid x = I, II, III \text{ and } y = I, II, III\}$. Since the selection is at random and both the patients and doctors are distinguishable among themselves, the uniform model seems appropriate.

In contrast, suppose you are asked to set up a sample space showing the number of patients each doctor saw when the patients were assigned at random to the doctors. Is the same sample space appropriate? We may wish to think of the sample space as being represented by three distinguishable boxes (the doctors) into which two distinguishable balls (the patients) are being tossed at random. The table below gives a visual representation of this new sample space S^* as well as a comparison of S and S^*.

Points in S^*	Boxes in S^*			Points in S
	I	II	III	
(AB, -, -)	A B	-	-	(I, I)
(-, AB, -)	-	A B	-	(II, II)
(-, -, AB)	-	-	A B	(III, III)
(A, B, -)	A	B	-	(I, II)
(A, -, B)	A	-	B	(I, III)
(-, A, B)	-	A	B	(II, III)
(B, A, -)	B	A	-	(II, I)
(-, B, A)	-	B	A	(III, II)
(B, -, A)	B	-	A	(III, I)

Note that as long as we consider the patients and doctors to be distinguishable among themselves there is a one-to-one correspondence between the points in S and S^*.

Now consider the question: How many patients were seen by each doctor? Here we are interested only in the number (not the identity) of the patients seen by each doctor. Although the patients are distinguishable, we will no longer treat them as such. In this case the points (A, B, -) and (B, A, -) in S^* represent the same outcome: Doctor I and doctor II each sees one patient. In terms of the number of patients seen by each doctor, the sample space can be rewritten as

$$S^{**} = \{(2, 0, 0), (0, 2, 0), (0, 0, 2), (1, 1, 0), (1, 0, 1), (0, 1, 1)\}.$$

In general terms we are interested only in the number of balls in each of the three boxes--the occupancy number.

Points in S^*	Points in S^{**}
(AB, -, -)	(2, 0, 0)
(-, AB, -)	(0, 2, 0)
(-, -, AB)	(0, 0, 2)
(A, B, -)	(1, 1, 0)
(B, A, -)	
(-, A, B)	(0, 1, 0)
(-, B, A)	
(A, -, B)	(1, 0, 1)
(B, -, A)	

Recall that each point in S^* had probability 1/9. Considering the relationship between points S^* and S^{**}, it is quite clear that

$$\Pr\{(2,\ 0,\ 0)\} = \Pr\{(0,\ 2,\ 0)\} = \Pr\{(0,\ 0,\ 2)\} = \frac{1}{9}$$

$$\Pr\{(1,\ 1,\ 0)\} = \Pr\{(1,\ 0,\ 1)\} = \Pr\{(0,\ 1,\ 1)\} = \frac{2}{9}$$

If the clinic administrators are interested only in the number of patients seen by a doctor, not necessarily a particular doctor, then the doctors, as well as the patients, can be considered as being indistinguishable. Then there are just two outcomes: (2 0 0) and (1 1 0). The unordered outcome (2 0 0) is composed of the ordered outcomes: (2, 0, 0), (0, 2, 0) or (0, 0, 2). Similarly, the outcome (1 1 0) is composed of (1, 1, 0), (1, 0, 1) or (0, 1, 1). Thus the probabilities of these unordered outcomes are

$$\Pr\{(2\quad 0\quad 0)\} = \frac{3}{9}$$

$$\Pr\{(1\quad 1\quad 0)\} = \frac{6}{9}$$

Example 12 further illustrates occupancy problems.

Example 12: Eight commuters drive cars to the city daily and park at random in one of three lots: A, B or C. (a) On a given day what is the probability that 5 cars will be parking in A, 2 will be in B and 1 will be in C? (b) Suppose we are interested in a 5, 2, 1 split but not necessarily to lot A, B and C in that order. How is the probability affected?

Solution: (a) Assuming the drivers choose the lots at random, there are 3^8 equally likely points in S. The eight cars can be partitioned into three groups so that 5 are in lot A, 2 in lot B and 1 in lot C in $8!/(5!2!1!)$ ways. Thus

$$\Pr\{5, 2, 1 \text{ split in A, B, C}\} = \frac{8!/(5!2!1!)}{3^8} .$$

(b) We are now allowing for all the permutations of a 5, 2, 1 split; hence,

$$\Pr\{5, 2, 1 \text{ split}\} = \frac{3!8!/(5!2!1!)}{3^8} .$$

3-6 Random Sampling

In Chapter 1 we talked about obtaining information about a population by examining a *sample* or subset from that population. Our main aim will be to select a sample which is representative of the population. There are many ways in which we might select a sample from a population.

Suppose we wish to know the average height in a class of 30 students. We do not have the time or resources to measure each student. Instead we decide to choose three students and carefully measure their heights. Several possible methods of selection come to mind.

1. Choose the first three students sitting in the first row.
2. Choose the first three poeple on alphabetical class register.
3. Allow instructor to choose three students.
4. Select three students at random.

We can quickly dispense with the first three methods of selection since they may lead to samples which do not accurately describe the population. Since it is very likely that short people tend to sit in the front of the class, method 1 would be biased toward short people. At first consideration method 2 may appear to be an acceptable procedure; however, it may contain a hidden bias. We do not know, but it is possible that people whose family names begin with the letters at the initial portion of the alphabet tend to have the same height. We can disregard method 3 on the basis that most individuals have personal biases which may unduly influence their selections. For instance, the instructor may try too hard to obtain the best "picture" of the average height by choosing three people: one short, one of medium height, one tall. But if two-thirds of the class is actually tall, this sample would not be representative.

Most of us would probably agree that method 4 is the best. Why? We have been conditioned to feel that selection at random ensures that each segment of the population is properly represented. We shall see that random sampling insures that each member of the population has an equal probability of being selected. Essentially random sampling is an application of the uniform probability model. Using the concepts of the uniform model we define a *random sample*.

Definition 3·4: *A sample will be called a* random sample *if all possible samples of this particular size (chosen under some specified selection scheme) have the same probability of being chosen.*

If we wish to adopt method 4, how shall we select the sample? Our previous discussion suggests three possible schemes for selection, all of which lead to a random sample:

1. Selection with replacement (WR) in which order is considered.
2. Selection without replacement (WOR) in which order is considered.
3. Selection without replacement (WOR) in which order is not considered.

Applying the counting techniques of Sections 3-2 and 3-3, we know that the number of points in the sample space for each possible selection is

$$(1)\ 30^3 \qquad (2)\ {}_{30}P_3 \qquad (3)\ {}_{30}C_3$$

In general if we had a population of N items from which a sample of size s is selected, then for each of the sampling schemes there are

$$(1)\ N^s \qquad (2)\ {}_{N}P_s \qquad (3)\ {}_{N}C_s$$

points in S.

Only in the first selection scheme is repetition permitted. In most practical problems one would not be interested in repetitions since they do not provide any additional information about the population. Hence, for most of our discussion we will restrict ourselves to sampling without replacement (WOR).

How shall we actually draw the sample? If we have 30 people and wish to choose only three for a committee, it should be immediately obvious that the order in which we choose the members is of no consequence. There are ${}_{30}C_3 = 4060$ subsets or samples. We could indicate the three members of each of the possible subsets on 4,060 slips of papers and then choose one slip of paper. Thinking that this procedure involves too much work, the ingenious student will suggest that we write down the names of the people in the population on 30 slips of paper and choose three of these, one after the other, and then neglect the order.

In what way are these two selection procedures equivalent? We know that the two schemes do not result in the same number of possible samples. The question we are interested in is: Will the probability of an unordered sample be the same under both schemes?

> Theorem 3·6: *Under the assumption of the uniform probability model, the probability associated with an unordered sample is the same under both ordered and unordered sampling without replacement.*

Proof:

Consider the ordered selection scheme in which the sampling is WOR. We know that there are ${}_NP_s$ possible ordered samples, which are equally likely assuming the uniform model. From our discussion of permutations and combinatons we know that each of s! of these has the same membership. Therefore, in the model for ordered sampling, each unordered sample has the same probability $s!/{}_NP_s$. We know that $s!/{}_NP_s = 1/{}_NC_s$ which is the probability associated with each unordered sample using the unordered model. □

Theorem 3·6 shows that to obtain a random sample of size s from a population of size N, we may first draw an ordered random sample and then disregard the order.

To illustrate another important property of the ordered WOR sample, let us reconsider the example of Section 3-1. Recall the class of three in which each student was called upon once. We found that

$$\Pr(\text{John must answer first}) = \frac{1}{3}$$

$$\Pr(\text{John must answer second}) = \frac{1}{3}$$

$$\Pr(\text{John must answer third}) = \frac{1}{3}$$

Similarly in the generalized example we can show that when three students are chosen from five the probability of a student, say John, being chosen on any particular draw is 1/5. This property is sometimes referred to as the "equivalence law of ordered sampling."

Theorem 3·7: *If a random ordered sample of size* s *is drawn from a population of size* N, *then on any particular one of the* s *draws each of the* N *items has the same probability* 1/N *of appearing.*

Proof:

To fix our ideas, suppose we are interested in drawing a particular item A on the rth draw. There are ${}_{N}P_{s}$ ordered samples. If A must be in the rth position, there are N - 1 items to fill the remaining s - 1 positions. Thus the remaining places can be filled in ${}_{N-1}P_{s-1}$ ways. The desired probability is then

$$\frac{{}_{N-1}P_{s-1}}{{}_{N}P_{s}} = \frac{(N - 1)(N - 2) \cdots (N - s + 1)}{N(N - 1)(N - 2) \cdots (N - s + 1)} = \frac{1}{N} . \qquad \square$$

As a consequence of this theorem, John should be prepared to answer each of the three questions. He has a 1 in 5 chance of being chosen for each!

Since there are only three questions, not all five people can be chosen to recite. What is the probability that John will be chosen?

> Theorem 3·8: *The probability that a specified item is included in a sample of size* s *taken at random from a population of* N *is* s/N.

Proof:

Here one particular item of the N must be included. In addition (s - 1) items must be chosen from the remaining (N - 1) items. If we are not interested in the order in which the items are chosen, the selection can take place in ${}_{N-1}C_{s-1}$ ways. Therefore, the probability of inclusion of a specified item is

$$\frac{\binom{N-1}{s-1}}{\binom{N}{s}} = \frac{s}{N} .$$

If the selection is ordered, the probability of inclusion remains s/N. For this case the probability of including a particular item on each of s mutually exclusive draws is 1/N. Thus the probability of inclusion of a specified item is

$$\sum_{i=1}^{s} \frac{1}{N} = \frac{s}{N} . \qquad \square$$

An examination of the enumerated sample space for the example will show that John (or any other member of the class) is included in six of the 10 possible samples. The probability of his inclusion is 3/5 as given by Theorem 3·7.

In drawing samples *with replacement*, we face the problem of the same item being selected more than once; that is, the problem of a "repetition."

Example 13: A child is playing with five blocks, all of which are different in color. He randomly selects one from his toy box and then replaces it. He repeats this operation five times. Describe an appropriate probability model for this experiment. What is the probability that he takes out (a) all different colored blocks? (b) Exactly four different colored blocks? (c) Exactly three different colored blocks?

Solution: A description of the probability model includes the number of points in S as well as the probability associated with each point. Here S consists of $5^5 = 3125$ ordered 5-tuples. Since the draws are at random, each point in S has probability 1/3125.

(a) This follows directly from fundamental counting principle:

$$\Pr(5 \text{ different colored blocks}) = \frac{5 \cdot 4 \cdot 3 \cdot 2 \cdot 1}{5^5} = \frac{24}{625} .$$

(b) Here one block is to be drawn twice: a repeater. The repeater can be chosen in five ways. After selecting the repeater, there are four different colors from which three are to be chosen with no repetition. This can be done in $\binom{4}{3}$ ways. We have now determined the composition of the event. But since order has been considered in the construction of S, we must also order the repeaters (two blocks) and the three different blocks. If [as in (a)] the five blocks were all different, they can be ordered in 5! ways. Due to the repeaters, only 5!/2! of these permutations are distinct. Hence the probability is

$$\frac{5!}{2!}\binom{4}{3}\frac{5}{5^5}.$$

(c) The event "three different colored blocks" may arise in two mutually exclusive ways: (A) A repeater may be chosen three times in conjunction with two different blocks <u>or</u> (B) Two blocks may each appear twice in conjunction with one different block. First consider Pr(A). As in part (b), the repeater can be selected in five ways and the remaining different blocks in $\binom{4}{2}$ ways. The five can be ordered in 5!/3! ways giving

$$\Pr(A) = 5\binom{4}{2}\frac{5!/3!}{5^5}.$$

By a similar argument the repeating pair can be selected in $\binom{5}{2}$ ways and the different block in three ways. Then, considering all the possible orderings, we have

$$\Pr(B) = \binom{5}{2}3\,\frac{5!/2!2!}{5^5}.$$

Since the events A and B are m.e., we have

$$\Pr(\text{exactly 3 different colors}) = \Pr(A) + \Pr(B) = \frac{300}{625}.$$

Using similar counting procedures we can find the probability of the events "exactly two different colors" and "all the same color." Note that these five events: exactly 5, 4, 3, 2, 1 different colors form a partition of S; hence, the sum of their probabilities is 1. Determination of these remaining probabilities is left as an exercise for the student.

In Examples 14 and 15 this problem is generalized to obtain the probabilities of repetitions in random sampling with replacement.

Example 14: A sample of size 3 is taken with replacement from a population of N items. (a) Discuss an appropriate sample space for this problem. (b) What is the probability that all three items chosen are different? (c) What is the probability that we select one item twice and the third different? (d) What is the probability that all three items are the same?

Solution: (a) The sample space has N^3 ordered outcomes. Since the sampling is at random, each outcome is equally likely. (b) We can select three different items in $N(N - 1)(N - 2) = {}_NP_3$ ways; hence,

$$\Pr(3 \text{ different items}) = \frac{{}_NP_3}{N^3} = \frac{(N - 1)(N - 2)}{N^2} .$$

Note that this is similar to the birthday problem discussed in Example 11. (c) Let us symbolize this result as R, R, D in any order. Then there are N choices for R, the repeater, and $(N - 1)$ choices for D. Since the sample space consists of ordered triplets we must count the number of ways the three items R, R, D can be ordered: $3!/(2!1!)$. Thus

$$\Pr(2 \text{ of one kind and } 1 \text{ different}) = \frac{3(N - 1)}{N^2} .$$

(d) Again the repeater can be chosen in N ways. Since the other two selections must be the same as the first,

$$\Pr(\text{all } 3 \text{ items are the same}) = \frac{1}{N^2} .$$

Example 15: Now suppose we select a sample of size s with replacement from a population of N items. Consider the same questions as posed in Example 14.

Solution: (a) The sample space consists of N^s equally likely outcomes.

(b)

$$\Pr(\text{all different}) = \frac{{}_{N}P_{s}}{N^{s}}$$

(c)

$$\Pr(2 \text{ of one kind and } s - 2 \text{ all different})$$
$$= \frac{(N - 1)^{C}(s - 2}{N^{s-1}} \frac{s!}{2!}$$

(d)

$$\Pr(\text{all the same}) = \frac{1}{N^{s-1}}$$

Detailed solutions are left as an exercise for the student.

Problems

1. Evaluate: (a) ${}_{9}P_{4}$, (b) ${}_{6}P_{6}$, (c) ${}_{n}P_{2}$, (d) ${}_{k}P_{1}$.

2. Evaluate: (a) ${}_{10}C_{3}$, (b) ${}_{6}C_{6}$, (c) ${}_{m}C_{2}$, (d) ${}_{t}C_{1}$, (e) ${}_{500}C_{498}$, (f) ${}_{k}C_{k-1}$.

3. Compute ${}_{n}P_{0}$ and interpret this result.

4. Compute ${}_{n}C_{0}$ and interpret this result.

5. Solve the following equations for n:

   (a) $\dfrac{{}_{n}P_{4}}{{}_{n}P_{2}} = 42$ (b) $\dbinom{n}{2} = 55$ (c) $\dfrac{{}_{n}P_{4}}{\dbinom{n-1}{3}} = 60$

6. Show that $\dbinom{n}{r} = \dbinom{n}{n-r}$. Interpret this result.

7. There are five roads from town A to town B, four roads from town B to town C and two directly from A to C.

   (a) How many ways can one make a trip from

      (i) A to C?

      (ii) A to C by way of B?

   (b) If a route from A to C is chosen at random, what is the probability that it goes through B?

8. Using only numerals from the set {1, 2, 3, 4} and with no repetitions, a list of all possible three digit numbers was formed. Each number was written on a cardboard disc. From these discs one disc was drawn at random.

8. (a) Discuss an appropriate sample space and probability model for this experiment.
   (b) What is the probability that the number on the selected disc
      (i) begins with the digit 1?
      (ii) begins with an even digit?
      (iii) is an odd number?

9. Repeat Problem 8 if repetitions are allowed.

10. A set of books contains eight volumes.
   (a) In how many ways can they be arranged on a shelf?
   (b) Suppose that three of the eight are written by author X and five by author Y. If they are now arranged at random, what is the probability that all the books by the same author are together?

11. Find the number of ways in which at least one flower can be selected from four (indistinguishable) roses and eight (indistinguishable) violets.

12. Answer Problem 11 if all 12 flowers are different.

13. If we pick two gloves at random from a drawer containing nine different pairs of gloves, what is the probability that
   (a) the two chosen are a right-hand and a left-hand but do not form a pair?
   (b) both are right hands?
   (c) a pair is chosen?

14. An agricultural experiment consists of comparing the yield of five varieties of wheat using three kinds of fertilizer and four types of insecticide. If a farmer has 64 plots, is this sufficient for comparing all possible combinations of varieties, insecticides and fertilizers?

15. Twenty players are entered in a badminton tournament. How many singles matches are necessary in order for each person to play each of the others exactly once?

16. A mouse in a maze has eight positions at which it must make a decision. The decision consists of going right or left.
    (a) In how many ways can the mouse run through the maze?
    (b) Suppose the reward (cheese) will be obtained if it always turns left at the last position. What is the probability the mouse gets the reward?

17. A package of assorted candies contains 14 in all of which five are cherry and three each are orange, pineapple and lime. In how many different ways can these candies be arranged in a row in a package?

18. How many four-letter "words" can be made from eight different letters
    (a) if any letter may be repeated any number of times?
    (b) if repetitions of a letter are not allowed?

19. Six people are chosen by lot from a group of 15 people to fill six distinct positions. If we label the 15 people from 1 through 15
    (a) what is the probability that person 1 is selected?
    (b) what is the probability that persons 4, 5, 6, 7, 14 and 15 are selected?

20. (a) How many arrangements are there of the letters of the word FEEBLENESS?
    (b) What is the probability that if the letters are arranged at random four E's will be together?
    (c) In a random arrangement, what is the probability that exactly three E's will be together?

21. Three boys and four girls are lined up at random in a row for a picture.
    (a) What is the probability that the girls and boys will alternate?
    (b) What is the chance that the girls are lined up in ascending order of height and the boys likewise?

22. Keys are made by placing notches of various depths at prescribed positions. Suppose each blank key has six positions at which it may or may not be notched. There are two cutting depths at the first position and three cutting depths at each of the remaining five positions. How many possible keys can be made?

23. A child's set of blocks consists of two red, four blue and five yellow cubes. The blocks can be distinguished only by color. If the child lines the blocks in a row at random, what is the probability that
    (a) a red block comes at both ends?
    (b) the five yellow blocks are together?
    (c) a blue block comes at both ends?

24. A group of nine children is to be used to evaluate two methods (A and B) of teaching reading. Four are selected at random to be taught by method A; the remaining five are assigned to method B. Find the probability that
    (a) the four most intelligent children are assigned to method A.
    (b) the three most intelligent children are assigned to method A.
    (c) the most intelligent child is assigned to method A.

25. Suppose that the group of nine students in Problem 24 consists of five boys and four girls. Two boys and two girls are selected at random for method A.
    (a) How many possible samples are there with two boys and two girls instructed by method A?
    (b) What is the probability that the most intelligent boy and most intelligent girl are in the group instructed by method A?

26. A square field is divided into 16 square plots of four rows and four columns. If three are chosen at random, what is the probability that:

(a) all the plots are in the same row?

(b) all the plots are in the same column?

(c) all are corner plots?

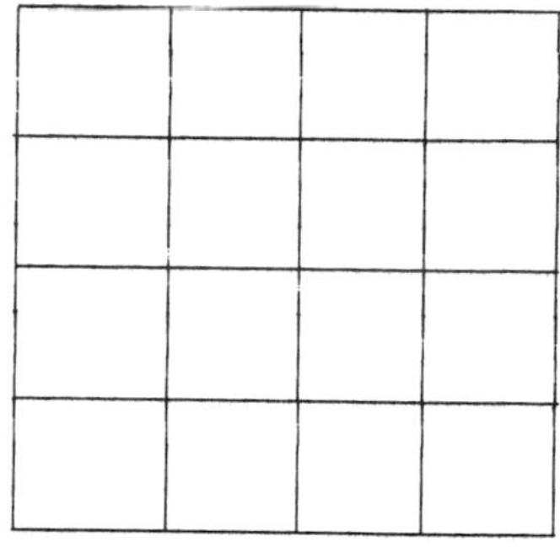

27. The weekend edition of a newspaper printed a recent photograph and a baby picture of three Canadian prime ministers. The reader was asked to match the recent and baby photographs of each. What is the probability that purely by chance the reader will get all three correct?

28. Two awards are to be distributed at random among 14 boys and 11 girls. Find the probability that
    (a) both awards will be given to boys.
    (b) both awards will be given to girls.
    (c) one award will be given to a girl and one to a boy.

29. If the awards in Problem 28 are a first and a second prize, what is the probability that a boy gets the first prize and a girl gets the second? Is this answer the same as that found for part (c) in Problem 28? Why or why not?

30. Of 10 girls in a class, four have blue eyes. If two girls are chosen at random (WOR), what is the probability that
    (a) both have blue eyes?
    (b) neither has blue eyes?
    (c) at least one has blue eyes?

31. Repeat Problem 30 if the selection is WR.

32. A population of a city block consists of 50 people: 25 liberals and 25 conservatives. A random sample of two people is chosen.
    (a) If the sampling is WOR, what is the probability that both are liberals?

(b) If the sampling is WR, what is the probability that both are liberals?

(c) If the sampling is WOR, what is the probability that a particular person, say Mr. X, is included in the sample?

33. Consider a population of N objects from which a random sample (WOR) of r items is taken. Find the probability that a specified r - 1 items will be included in the sample.

34. From a class of 100 students the instructor selects 10 people at random one after another (WOR) to answer 10 questions.

(a) What is the probability that Bob, a member of the class, is called upon to answer the seventh question?

(b) What is the probability that Bob is called upon to answer?

35. In a group of three people, what is the probability that all three people were born on different days of the week?

36. A box contains N distinct blocks labeled 1 to N. If k blocks are selected at random (WR), what is the probability that the same block is not drawn more than once?

37. An urn contains three red and four blue balls. A ball is selected at random. Its color is recorded and the ball is returned to the urn. This process is repeated six times.

(a) Discuss an appropriate sample space for this experiment.

(b) Does a uniform probability model seem appropriate? Why or why not?

38. If a balanced die is rolled k times, discuss an appropriate probability model for the experiment.

39. A box contains k fuses of which g are good and d are defective. If the fuses are drawn one by one (WOR) from the box and tested, what is the probability that the first and the last are defective?

40. An ordinary deck of 52 playing cards is shuffled and the cards are turned up one at a time until an ace appears. What is the probability that the first ace appears
    (a) on the sixth card?
    (b) on the forty-seventh card?
    (c) on the rth card?
    (d) at the rth or sooner?

41. Thirteen pieces of candy are to be distributed to five children. In how many ways can this be done if the first child must receive four pieces, the second child three pieces and the remaining children two pieces each?

42. In a carnival game 10 balls are thrown into three boxes. What is the probability that three balls are in one box, three balls in another box and four balls in a third box?

43. Four married couples attend a dance. For the first dance, one man and then one woman are chosen at random.
    (a) Construct a sample space to describe the first dancers and assign appropriate probabilities.
    (b) Consider the events:

        E: The two first dancers are married.
        F: Mr. A is chosen for the first dance.

        (i) Express E and F as subsets of the sample space and calculate their probabilities.
        (ii) Describe in words the event $E \cap \bar{F}$.

44. Suppose a field has eight plots arranged in two rows of four plots each. Two plots are to be chosen for an experiment. We are interested in the event E that both plots chosen are corner plots. Find Pr(E) under the following three methods of choosing the plots:
    (a) The two plots are chosen at random.

(b) One plot is chosen at random from each row.

(c) Two columns are chosen at random and then a plot is chosen at random from each of the chosen columns.

45. Three cards are dealt from a well shuffled deck of four red and 12 black cards. Find the probability that the third card dealt is red.

Chapter 4

Conditional Probability

Frequently we may be interested in probabilities concerning part, rather than all, of a sample space. From research in genetics it is known that hair and eye coloring are associated. Due to this association, the probability that a person chosen at random from a population has blue eyes will be different from the probability of blue eyes in the subpopulation of people with blonde hair.

Here we are concerned with extra conditions imposed by a subset of the population which may not apply to the population-at-large. Probabilities associated with these subpopulations are called *conditional probabilities*.

Conditional probability is also useful if during an experiment the conditions are altered so that partial information concerning the outcomes becomes available. For example, if a balanced die is rolled, the probability of a "1" is 1/6. If, however, you are given the added information that the outcome is odd, the probability of a "1" becomes 1/3. In this situation the sample space of the experiment has been altered.

4-1 Reduction of the Sample Space

In order to investigate further how added information about the outcomes of an experiment affect the sample space, let us consider the following example.

Example 1: Suppose a family is chosen at random from a set of families each having exactly two children (not twins). What is the probability that both children are boys? Quite reasonably we could adopt a sample space of equally likely ordered pairs:

$$S = \{(B, B), (G, B), (B, G), (G, G)\}.$$

Then

$$\Pr(\text{both children are boys}) = \frac{1}{4}.$$

Now you learn that the data have been obtained from two-children families in a boys' school. How does this affect the probability? We would all agree that a possible new sample space is

$$S^* = \{(B, B), (G, B), (B, G)\}.$$

Are these outcomes equally likely? We are now looking for the probability of two boys when we know that at least one is a boy.

Formally we would denote this probability by

$$\Pr\{\text{two boys} \mid \text{at least one boy}\}$$

where the vertical line is read "given that" and should not be confused with the division symbol /.

How shall we assign probabilities to the outcomes in S^*? In S the event "at least one boy" has probability 3/4. Since we now want the $\Pr(S^*)$ to be 1, we must enlarge the probabilities of the outcomes in S^* by dividing the probability of each outcome by the probability of the conditioning event. Hence

$$\Pr\{\text{two boys} \mid \text{at least one boy}\} = \frac{1/4}{3/4} = \frac{1}{3}.$$

In general consider an arbitrary sample space $S = \{E_1, E_2, \cdots, E_k\}$ with the events A, B and their intersection $A \cap B$.

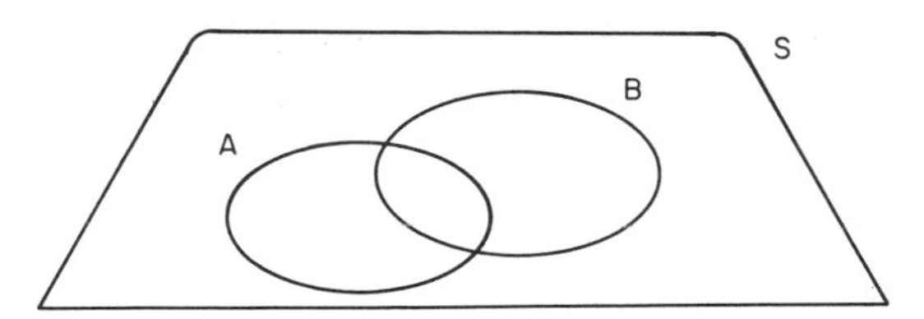

Now suppose we are interested in conditioning on the event B. If we are given B, then we ignore all other outcomes in S and think of B as the new *reduced* sample space, S^*.

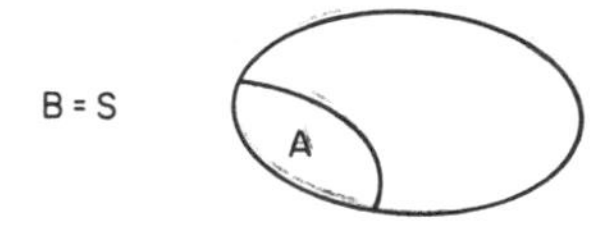

Since we know that B has occurred, the number of possible outcomes has been reduced. Thus the probability associated with each elementary event in S^* must be increased so that the total probability of S^* now equals 1. In defining these new probabilities we will wish to maintain the relative relationship of the probabilities among themselves; that is, if $Pr(E_i)$ was twice $Pr(E_j)$ in S, then we want this same relationship to exist in S^*. Let $Pr^*(E_i)$ denote the new probability of the event E_i for those elementary events contained in S^*. To maintain the relative relationship of their probabilities, we will multiply all the $Pr(E_i)$ in S^* by a constant of proportionality:

$$Pr^*(E_i) = cPr(E_i).$$

To determine c we note that $\Sigma\ Pr^*(E_i)$ summed over the E_i in S^* must equal one. Hence

$$c\ \Sigma\ Pr(E_i) = 1.$$

with the summation over all E_i in S^*. Solving for c, we get

$$c = \frac{1}{\Sigma\ Pr(E_i)}.$$

Since B is the union of E_i in S^*,

$$c = \frac{1}{Pr(B)}.$$

Thus for S^* to constitute a sample space, each elementary event must have its probability enlarged by $1/Pr(B)$.

We have discussed how each point in the reduced sample space can be scaled to maintain a proper probability model. How does this affect the probability of a particular event A? In Example 1,

$$A = \{\text{two boys}\}$$
$$B = \{\text{at least one boy}\}.$$

Since all the elements of S, except those in B, have been eliminated, the only elements of A which remain are those in $A \cap B$. From the basic definition of probability, we would determine the probability of A relative to the new sample space $S^* = B$ by summing the $Pr^*(E_i)$ for all E_i contained in A:

$$Pr(A \mid B) = \sum_{\text{all } E_i \text{ in } A} Pr^*(E_i).$$

Is it possible to determine this probability from the probabilities associated with the original sample space?

Recall that

$$Pr^*(E_i) = \frac{Pr(E_i)}{Pr(B)} ;$$

hence,

$$Pr(A \mid B) = \sum_{\text{all } E_i \text{ in } A} Pr^*(E_i) = \sum_{\text{all } E_i \text{ in } A} \frac{Pr(E_i)}{Pr(B)}$$

$$= \frac{1}{Pr(B)} \sum_{\text{all } E_i \text{ in } A} Pr(E_i).$$

How else can the expression

$$\sum_{\text{all } E_i \text{ in } A} Pr(E_i)$$

be written? This is just the probability of the intersection $(A \cap B)$ with respect to the original sample space S; thus

$$Pr(A \mid B) = \frac{Pr(A \cap B)}{Pr(B)} .$$

In most applications we will find that it is not necessary to go through the complicated procedure of finding the probability associated with each outcome in the reduced sample space. Rather we can work directly with the probabilities of the original sample space.

> Definition 4·1: *The* conditional probability *of* A, *given* B, *is denoted by* Pr(A | B) *and is defined by*
>
> $$\Pr(A \mid B) = \frac{\Pr(A \cap B)}{\Pr(B)} \quad \textit{for } \Pr(B) \neq 0.$$

Actually all probabilities are conditional probabilities in that they refer to some sample space. We use the abbreviated symbol Pr(B) when we really mean Pr(B | S). If, however, we are considering a particular subset of S, it becomes necessary to state the conditioning event explicitly.

We shall see that the general theorems concerning probability developed in Chapter 2 are also valid for conditional probabilities.

Let us now illustrate the basic concepts of conditional probability by considering the following examples.

Example 2: A box contains four balls labeled A, B, C, D. Two balls are drawn one after the other with replacement. Given that the same ball does not appear on both draws, what is the probability that neither of the two balls is B?

Solution: The unrestricted sample space consists of 16 possible outcomes. Under the assumption of random sampling each outcome is equally likely.

		First draw			
		A	B	C	D
Second draw	A	A,A	B,A	C,A	D,A
	B	A,B	B,B	C,B	D,B
	C	A,C	B,C	C,C	D,C
	D	A,D	B,D	C,D	D,D

Now suppose we wish to reduce the sample space by conditioning on the event "same ball does not appear on both draws." The reduced sample space S^* does not contain the outcomes lying on the diagonal. Each outcome in S^* has probability: $(1/16)/(3/4) = 1/12$. By examination of S^*, we see that the conditional probability

$$\Pr(\text{neither is B} \mid \text{not two the same}) = \frac{6}{12} = \frac{1}{2} .$$

We could also obtain this probability by using the original sample space S and the definition of conditional probability:

$$\Pr(\text{neither is B} \mid \text{no two the same})$$

$$= \frac{\Pr(\text{neither is B} \cap \text{no two the same})}{\Pr(\text{no two the same})} = \frac{6/16}{12/16} = \frac{1}{2} .$$

Note that S^* is the sample space for ordered sampling WOR; that is, the restriction of "no repetition" in ordered sampling WR leads to ordered sampling WOR.

Example 3: Recently a sample survey was undertaken to investigate attitudes toward breakfast programs in elementary schools. The responses were classified as "favor", "against" or "no opinion." Since it was felt that the attitudes in the suburban areas may be different from those in the inner city, the 250 respondents were also classified according to their place of residence. The results of the survey are given in the following 2 × 3 table:

Place of Residence	*Attitude Toward Breakfast Program*		
	Favor	*Against*	*No Opinion*
Inner City	100	30	15
Suburbs	50	50	5

If an individual is selected at random from these 250, find the probability that

(a) he lives in the inner city.

(b) he lives in the inner city and favors the program.

(c) he favors the program if you know that he lives in the inner city.

(d) he lives in the inner city if he favors the program.

Solution: Let us represent several events of interest symbolically:

I: living in the inner city

S: living in the suburbs

F: favors the program

A: against the program

N: no opinion concerning the program.

(a) Since the events F, A and N are m.e., the event I can be written as

$$I = (I \cap F) \cup (I \cap A) \cup (I \cap N).$$

Application of the addition law gives

$$\Pr(I) = \Pr(I \cap F) + \Pr(I \cap A) + \Pr(I \cap N)$$

$$= \frac{100}{250} + \frac{30}{250} + \frac{15}{250} = \frac{145}{250}.$$

(b) Reading directly from the table, we have $\Pr(I \cap F) = 100/250$.

(c) Combining the results of parts (a) and (b) and using the definition of conditional probability, we get

$$\Pr(F \mid I) = \frac{\Pr(I \cap F)}{\Pr(I)} = \frac{(100/250)}{(145/250)} = \frac{100}{145}.$$

This probability can also be determined directly by considering the subset of the survey living in the inner city. Of these 145 people, 100 favor the program; hence,

$$\Pr(F \mid I) = \frac{100}{145}.$$

(d) We can again approach this problem from a subset point of view. Of the 150 people who favor the program, 100 are from the inner city:

$$\Pr(I \mid F) = \frac{100}{150}.$$

How might this question be answered using the formal definition of conditional probability? Definition 4·1 gives

$$\Pr(I \mid F) = \frac{\Pr(I \cap F)}{\Pr(F)}.$$

As in part (a) the event

$$F = (F \cap I) \cup (F \cap S)$$

with

$$\Pr(F) = \Pr(F \cap I) + \Pr(F \cap S)$$

$$= \frac{100}{250} + \frac{50}{250} = \frac{150}{250}.$$

Then

$$\Pr(I \mid F) = \frac{(100/250)}{(150/250)} = \frac{100}{150}.$$

4-2 Multiplication Rule And Assigning Probabilities

From Definition 4·1 we know that

$$\Pr(A \mid B) = \frac{\Pr(A \cap B)}{\Pr(B)} \quad \text{for } \Pr(B) \neq 0.$$

If we multiply both sides by Pr(B), then

$$\Pr(A \cap B) = \Pr(A \mid B)\Pr(B).$$

This *multiplication rule*, often called the *product law of probability*, is useful in assigning probabilities to the intersections of events.

Example 4: Assume that 5% of males (M) and 1% of females (F) are color-blind (C) and that males and females each form 50% of the population. (a) If a person is chosen at random from the population, what is the probability that he is a male? (b) If a person is chosen at random from the population of color-blind individuals, what is the probability that he is a male?

Solution: The four intersection probabilities $(M \cap C)$, $(M \cap N)$, $(F \cap C)$ and $(F \cap N)$ can be found by using the multiplication rule. The problem states that 5% of males and 1% of females are color-blind. These percentages represent the conditional probabilities of the events $C \mid M$ and $C \mid F$, respectively. Thus

$$\Pr(M \cap C) = \Pr(C \mid M)\Pr(M) = (.05)(.50) = .025$$
$$\Pr(F \cap C) = \Pr(C \mid F)\Pr(F) = (.01)(.50) = .005.$$

Since N and C are complementary events, we can easily write

$$\Pr(M \cap N) = \Pr(N \mid M)\Pr(M) = (.95)(.50) = .475$$
$$\Pr(F \cap N) = \Pr(N \mid F)\Pr(F) = (.99)(.50) = .495.$$

These results can be summarized in the following 2 × 2 table:

	Color-Blind (C)	*Normal Vision (N)*
Male (M)	.025	.475
Female (F)	.005	.495

The unconditional probability of being male is (as stated in the example) .50. This can also be obtained from the 2 × 2 table as

$$\Pr(M) = \Pr\{(M \cap C) \text{ or } (M \cap N)\}.$$

Since the events $(M \cap C)$ and $(M \cap N)$ are m.e.,

$$\Pr(M) = .025 + .475 = .50.$$

To find the conditional probability in the subpopulation of color blind individuals we use Definition 4·1:

$$\Pr(M \mid C) = \frac{\Pr(M \cap C)}{\Pr(C)}.$$

We determine Pr(C) as we did Pr(M):

$$\begin{aligned}\Pr(C) &= \Pr\{(M \cap C) \text{ or } (F \cap C)\} \\ &= .025 + .005 = .030.\end{aligned}$$

Therefore

$$\Pr(M \mid C) = \frac{.025}{.030} = \frac{5}{6}.$$

Example 5: Suppose a box contains 9 poker chips: 6 black and 3 red. Two chips are chosen at random one after the other without replacement. Set up an appropriate sample space and assign probabilities to the outcomes.

Solution: Since the chips can only be distinguished on the basis of color, a possible sample space is

$$S = \{(R, R), (R, B), (B, R), (B, B)\},$$

where, for instance, (R, B) means a red chip on the first draw followed by a black chip in the second. If we consider the outcomes in S as the intersections of the events "first draw outcome" and "second draw outcome," we can use the multiplication rule to assign the probabilities:

$$\begin{aligned}\Pr\{(R, R)\} &= \Pr(\text{red on first draw})\Pr(\text{red on second draw} \mid \text{red on first draw}) \\ &= \frac{3}{9}\frac{2}{8} = \frac{1}{12}.\end{aligned}$$

Similarly

$$\Pr\{(R, B)\} = \frac{3}{9}\frac{6}{8} = \frac{1}{4}$$

$$\Pr\{(B, R)\} = \frac{6}{9}\frac{3}{8} = \frac{1}{4}$$

$$\Pr\{(B, B)\} = \frac{6}{9}\frac{5}{8} = \frac{5}{12}.$$

Note that each of these outcomes is not equally likely.

The multiplication rule can easily be extended for any number of events; hence, for three events A, B, C we have

$$\Pr(A \cap B \cap C) = \Pr(A)\Pr(B \mid A)\Pr(C \mid A \cap B).$$

4-3 Stagewise Experiments

Many experiments can be decomposed into stages or trials. In Example 5 we were drawing two chips from a box. It may be useful to treat this experiment as having two related stages. The following tree diagram gives a pictorial view of the stages.

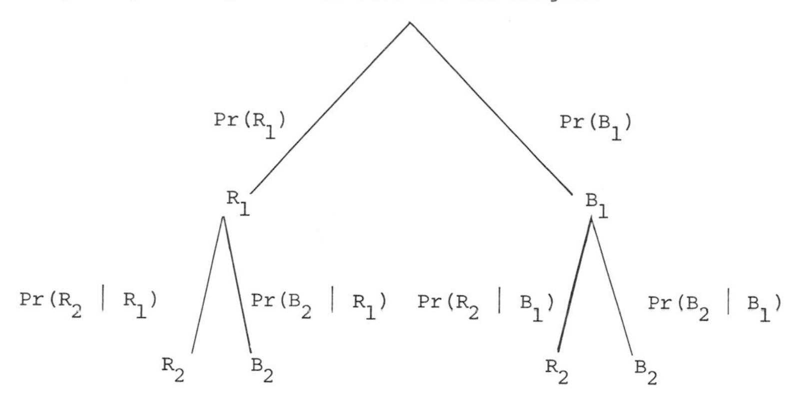

We label each branch with the appropriate probability. Since the stages (draws) are related, the probabilities at the second stage are affected by the outcome at the first stage. Hence, the second stage probabilities are conditional. To find the probability of an intersection we multiply the corresponding branch probabilities.

Suppose now we wish to find the probability that the sample will contain a red and a black chip. This can occur in two m.e. ways:

$$RB = R_1B_2 \cup B_1R_2.$$

To find the probability of this event, we add the probabilities of the inner branches of the tree diagram:

$$\begin{aligned} Pr(RB) &= Pr(R_1B_2) + Pr(B_1R_2) \\ &= Pr(R_1)Pr(B_2 \mid R_1) + Pr(B_1)Pr(R_2 \mid B_1) \\ &= \frac{1}{2} . \end{aligned}$$

We know that the probability of a red chip on the first draw is 1/3. What is the probability of a red chip in the second draw? From the equivalence law we know that this probability should remain 1/3. Does our stagewise development support this result?

The event "red on second draw" can be expressed as

$$R_2 = R_1R_2 \cup B_1R_2.$$

Again the events on the r.h.s. are m.e.; thus

$$\begin{aligned}\Pr(R_2) &= \Pr(R_1)\Pr(R_2 \mid R_1) + \Pr(B_1)\Pr(R_2 \mid B_1) \\ &= \frac{1}{12} + \frac{1}{4} = \frac{1}{3}\end{aligned}$$

as we expected.

Example 6: Suppose you are attending a meeting of the "Parents of Twins." If you choose at random a family which has just one pair of twins (no other children), what is the probability that both twins are boys? It will also be useful to know that approximately 1/3 of all human twins are identical.

Solution: This experiment can be thought of as having two stages: type of twins (identical I or fraternal Fr) and sex of twins (MF, FM, FF, MM). The stagewise outcomes and probabilities can be summarized by a tree diagram:

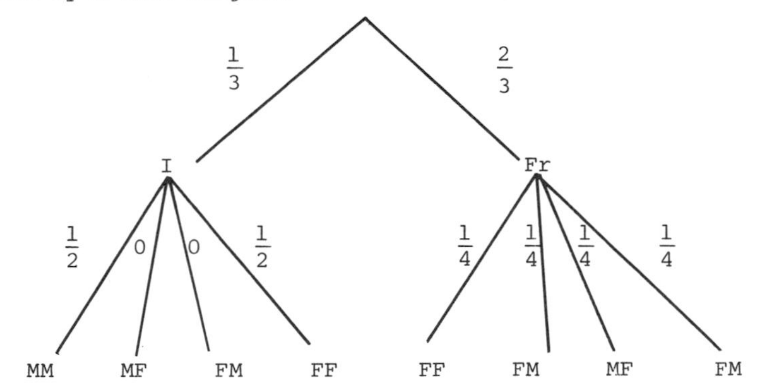

Notice that the sex pattern of the fraternal twins is like that of two-children families. In contrast, for identical twins the outcomes MF and FM are no longer possible. Essentially we are conditioning on the event "same sex" in families with two children.

Recalling that the second stage probabilities are conditional, we can write

$$\begin{aligned}\Pr(\text{twins are boys}) &= \Pr(I \cap MM) + \Pr(Fr \cap MM)\\ &= \Pr(MM \mid I)\Pr(I) + \Pr(MM \mid Fr)\Pr(Fr)\\ &= \frac{1}{2}\frac{1}{3} + \frac{1}{4}\frac{2}{3} = \frac{1}{3}\,.\end{aligned}$$

Recall that in two-children families Pr(MM) = 1/4. Suppose that you have undertaken a study in which you observe the relative frequency of MM in two children families. You find it is larger than 1/4. How might you explain this result? This inflated probability of MM might be due to the presence of twins in the study.

Example 7: A third year Arts student finds at the end of his final year that he is lacking a science course needed for graduation. He examines the summer school offerings and finds that he can take a course in mathematics (M), chemistry (C) or computer science (CS). On the basis of his interest he assigns a probability of .1, .6, .3 to the events of choosing each of these. After considering his past performance, his advisor estimates the probability of his passing (P) the mathematics course as .80, the chemistry course as .70 and the computer science course as .75. (a) What is the probability of his passing? (b) If at the end of the summer you hear that he has graduated, what is the probability that he took the (i) mathematics course, (ii) chemistry course, (iii) computer science course?
Solution: Again a tree diagram may be useful in summarizing the data.

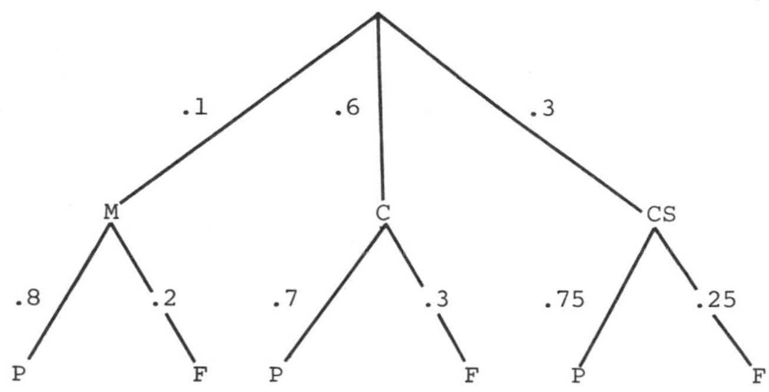

(a) $$\Pr(P) = \Pr(P \cap M \text{ or } P \cap C \text{ or } P \cap CS)$$

As before the three events on the r.h.s. are m.e.; hence, using the addition law and Definition 4·1 we have

$$\begin{aligned} \Pr(P) &= \Pr(P \mid M)\Pr(M) + \Pr(P \mid C)\Pr(C) + \Pr(P \mid CS)\Pr(CS) \\ &= (.8)(.1) + (.7)(.6) + (.75)(.3) = .725. \end{aligned}$$

(b) How would you attack this problem? The experiment has been completed; the student has passed. We are now looking for probabilities $\Pr(M \mid P)$, $\Pr(C \mid P)$ and $\Pr(CS \mid P)$. In the next section we will discuss how these probabilities can be found.

4-4 Posterior Probabilities: Bayes' Rule

In this section we shall use conditional probability to modify the probability of an event as a result of experimental evidence. Up to this point in our discussion, all probabilities have been determined prior to the actual execution of the experiment. In Example 7 we saw that prior to the experiment, the probability that the student would pass was .725. Now the summer is over and the student has graduated. He must have passed one of the courses! What is the probability that he took the mathematics course? Probabilities, determined after the completion of an experiment, are often referred to as *inverse* or *posterior probabilities*.

Applying the definition of conditional probability in Example 7, we see that

$$\Pr(M \mid P) = \frac{\Pr(M \cap P)}{\Pr(P)} .$$

We have already determined the denominator probability. Using the multiplication rule, we can write the numerator as $\Pr(P \mid M)\Pr(M)$; hence

$$\Pr(M \mid P) = \frac{\Pr(P \mid M)\Pr(M)}{\Pr(P)} = \frac{.08}{.725} = .1103.$$

If we refer to the tree diagram, we see that this posterior probability is the ratio of the branch resulting in passing by way of mathematics to all possible branches resulting in passing.

Our knowledge of the outcome of the experiment has permitted us to modify the probabilities:

Prior Probability	*Posterior Probability (P)*	*Posterior Probability (F)*
Pr(M) = .1	.1103	.0727
Pr(C) = .6	.5793	.6545
Pr(CS) = .3	.3103	.2727

Note that either outcome P or F gives more information and allows us to alter the original prior probabilities

This example illustrates a general proposition known as Bayes' Rule. Recall that in Chapter 2 we showed that any event F can be decomposed into the union of mutually exclusive components:

$$F = (A_1 \cap F) \cup (A_2 \cap F) \cup \cdots \cup (A_k \cap F),$$

where $A_1, A_2, \ldots, A_k$ are m.e. events each with nonzero probability and one of which must occur.

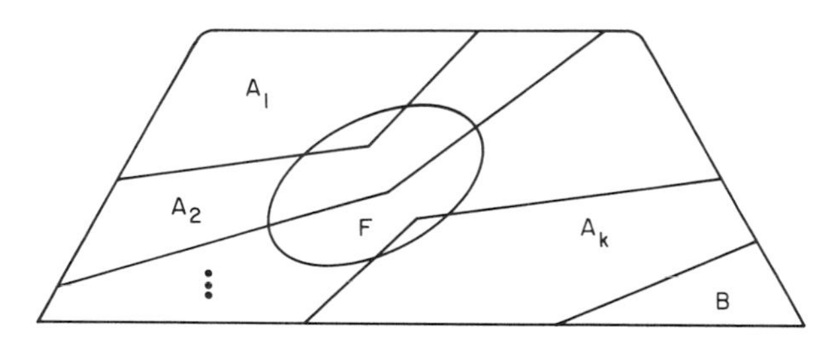

Note that the events $A_1, A_2, \ldots, A_k$ may, but need not necessarily form a partition of S. Applying the addition law for k mutually exclusive events, we have

$$\Pr(F) = \sum_{i=1}^{k} \Pr(A_i \cap F).$$

Now using the multiplication rule for the probability of each intersection, we find that

$$\Pr(F) = \sum_{i=1}^{k} \Pr(F \mid A_i)\Pr(A_i).$$

Suppose we know that F has occurred. From our diagram we see that it could have occurred with $A_1, A_2, \ldots, A_k$. What then is the probability that a particular event A_j occurred <u>given that we have observed</u> F?

$$\Pr(A_j \mid F) = \frac{\Pr(A_j \cap F)}{\Pr(F)}$$

$$= \frac{\Pr(F \mid A_j)\Pr(A_j)}{\sum_{i=1}^{k} \Pr(F \mid A_i)\Pr(A_i)}$$

for $j = 1, 2, \ldots, k.$

Formally we can summarize this development.

Bayes' Rule: *Let* F *be an arbitrary event in* S *such that* F *is a subset of the union of m.e. events* $A_1, A_2, \ldots, A_k$ *and* $\Pr(F) > 0$. *Then*

$$\Pr(A_j \mid F) = \frac{\Pr(F \mid A_j)\Pr(A_j)}{\sum_{i=1}^{k} \Pr(F \mid A_i)\Pr(A_i)} \quad \text{for } j = 1, 2, \ldots, k.$$

In the next example we will see how Bayes' Rule can be used in medical diagnostic testing.

Example 8: Often it is quite expensive in terms of money, personnel and time to carry out medical diagnostic tests on the entire population. In such cases screening tests are used. For example, it may not be possible to take X rays of the whole population in attempting to diagnose tuberculosis. Instead a skin test (BCG) is used as a preliminary screening device. The usual procedure is to follow up with an X ray all those who have a positive reaction the skin test.

The difficulty with these diagnostic tests is that they are never perfect. Two kinds of errors may arise:

1. False-positive: The test indicates the person has the disease when he does not.
2. False-negative: The test fails to indicate the disease even when the person has the disease.

We know that in the population-at-large a certain proportion has the disease (D). If a diagnostic procedure is applied, the reaction will be positive (+) or negative (-). Thus the probability of a false-positive is $\Pr(+ \mid \bar{D})$ and the probability of a false-negative is $\Pr(- \mid D)$.

Suppose we are interested in finding how many people out of 1000 will have to undergo more extensive clinical examination. From past experience we know that the prevalence of the disease in the population Pr(D) is .01, $\Pr(+ \mid \bar{D}) = .1$ and $\Pr(- \mid D) = .2$. The probability that a person tests positive is

$$\begin{aligned} \Pr(+) &= \Pr[(+ \cap D) \text{ or } (+ \cap \bar{D})] \\ &= \Pr(+ \mid D)\Pr(D) + \Pr(+ \mid \bar{D})\Pr(\bar{D}) \\ &= (.8)(.01) + (.1)(.99) = .107; \end{aligned}$$

hence about 11% of the people will have to be further examined.

Now let us consider the inverse probability. Suppose a person is declared positive; what is the probability that he actually has the disease? Using Bayes' Rule, we have

$$\Pr(D \mid +) = \frac{\Pr(+ \mid D)\Pr(D)}{\Pr(+)} = \frac{(.8)(.01)}{.107} = .075.$$

In this case about 92.5% of those who show a positive reaction on the screening test actually do not have the disease. Do you find this conclusion somewhat startling? Since the disease is quite rare (1% prevalence) and the rate of false-positives is only 10%, the bulk of the positive reactions must come from the healthy group.

Example 9 further illustrates the use of conditional probability in modifying probabilities.

Example 9: In an anthropological study an attempt is being made to classify a bone as belonging to one of two groups (I or II) of individuals. Suppose that a particular type of defect (E) is rare. We know that the probability that E occurs in group I is p (very small), while the probability of its occurring in II is p^2. Prior to our examination of the bone, we feel that it is equally likely to have come from either group. How is the classification of bone affected by (a) observing the rare event E? (b) failing to observe E?

Solution: Let us summarize the information given in the problem.

$$\Pr(E \mid I) = p \qquad \Pr(E \mid II) = p^2$$
$$\Pr(I) = \frac{1}{2} \qquad \Pr(II) = \frac{1}{2}$$

(a) Now if we observe E we can use Bayes' Rule to find

$$\Pr(I \mid E) = \frac{\frac{1}{2}p}{\frac{1}{2}p + \frac{1}{2}p^2} = \frac{1}{1+p}.$$

Since p is very small, $\Pr(I \mid E)$ is approximately 1. Similarly,

$$\Pr(II \mid E) = \frac{\frac{1}{2}p}{\frac{1}{2}p + \frac{1}{2}p^2} = \frac{p}{1+p},$$

which is approximately zero. Thus the occurrence of E makes classification to group I virtually certain.

(b) If, in contrast, we fail to observe E, we know that

$$\Pr(\bar{E} \mid I) = 1 - p \qquad \Pr(\bar{E} \mid II) = 1 - p^2.$$

Bayes' Rule gives the inverse probabilities

$$\Pr(I \mid \bar{E}) \quad \text{and} \quad \Pr(II \mid \bar{E})$$

each as approximately 1/2. In this case the failure to observe E is of little aid in the classification since the prior and posterior probabilities are the same.

One of the roles of statistics is decision making. Here we have seen how conditional probability can be used to modify hypotheses and assign adjusted weights (probabilities) to them after the experiment has taken place.

Prior to the experiment the probabilities associated with a set of n hypotheses may be

$$\Pr(H_1), \Pr(H_2), \ldots, \Pr(H_n).$$

Bayes' Rule then shows us how to use the information provided by the experiment and find posterior probabilities:

$$\Pr(H_1 \mid E), \Pr(H_2 \mid E), \ldots, \Pr(H_n \mid E).$$

Problems

1. Consider a sample space consisting of five simple events with probabilities $\Pr(E_1) = .2$, $\Pr(E_2) = .15$, $\Pr(E_3) = .15$, $\Pr(E_4) = .35$ and $\Pr(E_5) = .15$. Let the event $A = E_1 \cup E_3 \cup E_4$.
   (a) Find the conditional probability model using A as the conditioning event.
   (b) Verify that it is a proper probability model.
2. Recall the two-dice experiment of Problem 6 in Chapter 2.
   (a) If we are told that the sum of the points on the two dice is 4, construct the conditional probability model.
   (b) Find $\Pr(\text{number on green die} \leq 2 \mid \text{sum} = 4)$ using the conditional probability model of part (a).

2. (c) Determine Pr(number on green die $\leq 2 \mid$ sum = 4) using the definition of conditional probability. Compare your answer with that obtained in part (b).

3. Show that if the original probability model is uniform, the nonzero probabilities in any conditional model will also be uniform. Why is this result useful?

4. Again refer to the two-dice problem. Now use Definition 4·1 to determine the following probabilities:
   (a) Pr(sum < 6 | green shows 2)
   (b) Pr(sum = 11 | red shows 5)
   (c) How does the conditioning event in (a) and in (b) affect the probabilities of events in question?

5. A balanced coin is tossed until a head appears or until it has been tossed four times. Given that a head did not appear on either of the first two tosses, find the probability that
   (a) the coin was tossed four times.
   (b) the coin was tossed just three times.

6. Suppose Pr(A) = .2, Pr(B) = .4 and Pr(A ∪ B) = .5.
   (a) Construct a 2 × 2 table showing the probabilities of the events $(A \cap B)$, $(\bar{A} \cap \bar{B})$, $(\bar{A} \cap B)$ and $(A \cap \bar{B})$.
   (b) Determine the probabilities of the following events
   (i) $A \mid B$, (ii) $B \mid A$, (iii) $\bar{A} \mid \bar{B}$, (iv) $\bar{A} \cup \bar{B}$

7. A balanced coin is tossed three times:
   (a) Construct the conditional sample space for the conditioning event: "at least two heads."
   (b) Assign appropriate probabilities to each point in the conditional sample space.
   (c) How does this conditioning information affect the probability of three heads?

8. If the events A and B are mutually exclusive and Pr(B) ≠ 0, what can you say about Pr(A | B)? How would you interpret this result?

9. If B is a subset of A, what can you say about $Pr(A \mid B)$? How would you interpret this result?

10. In a tasting experiment four different blends of coffee A, B, C and D are given to a panel in a random order.
    (a) Define a sample space to describe the order of occurrence of the four blends.
    (b) Assign appropriate probabilities to each point in S.
    (c) Define the events: G: B is given last
    H: A is given before C
    Find the probabilities of the events:

(i) G	(iv) $H \cup G$	(vii) $H \mid \bar{G}$
(ii) H	(v) $G \mid H$	(viii) $\bar{G} \mid H$
(iii) $G \cap H$	(vi) $H \mid G$	

11. Prove that $Pr(A \cap B \cap C) = Pr(A)Pr(B \mid A)Pr(C \mid A \cap B)$.

12. Suppose a box contains b black chips and w white ones. If two chips are drawn at random without replacement, compute
    (a) Pr(second chip black | first chip white)
    (b) Pr(second chip white)
    (c) Pr(first chip black | second chip white).

13. Repeat Problem 12 if the sampling is with replacement.

14. A vaccination procedure calls for a first shot of vaccine to which the probability of an adverse reaction is .4. If there is a detectable adverse reaction, no further vaccine is given. If, however, there is no detectable reaction to this first shot, a second shot is given. For those receiving this second shot, the probability of a reaction is .6.
    (a) What is the probability that an individual selected at random will react only to the second dose of vaccine?
    (b) What is the probability that an individual selected at random will react to either shot of vaccine?

15. Three identical black balls are thrown at random into three boxes. Find the conditional probability that all three balls will be in the same box, given that at least two of them are in the same box.

16. In a learning experiment a mouse runs a T maze twice. The mouse has two choices on any run: (1) go to the left and get food or (2) go to the right and receive an electric shock. On the first run, the mouse is equally likely to go to the left or to the right. If he receives food on the first run, the probabilities that he goes to the left or to the right on the second run are .8 and .2, respectively. If he receives an electric shock on the first run, the probabilities that he goes to the left or to the right on the second run are .6 and .4, respectively.

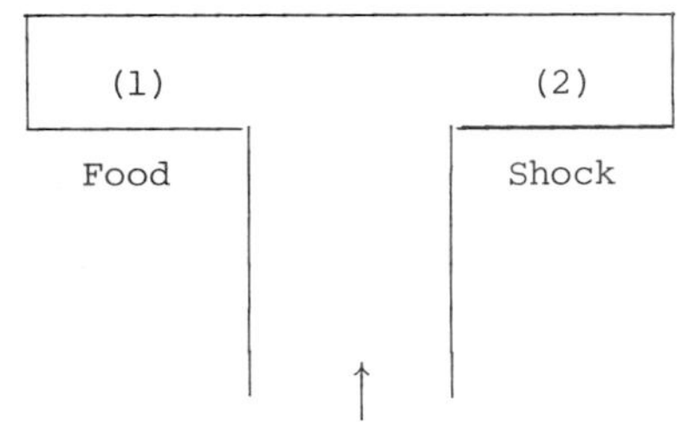

(a) Enumerate the possible outcomes after two runs and assign appropriate probabilities to each.

(b) Find the probability that the mouse goes to the right on the second run.

(c) If you observe the mouse at the right on the second run, what is the probability that he selected the right on the first run?

17. A drawer has four black socks, six brown socks and two blue ones. What is the probability that two socks drawn at random (WOR) are both the same color?

18. A history test is made up of five true-false-type questions. What is the probability that a student will get all answers correct if

(a) He is only guessing?

(b) He knows that the teacher puts more true than false questions on tests?

18. (c) In addition he knows that the teacher never puts three questions in a row with the same answer?
    (d) He <u>also</u> knows that the first and last questions must have opposite answers?
    (e) He knows in addition that the answer to the second question is "false?"

19. Suppose N psychiatrists interview a patient and are then asked to say whether or not he is a schizophrenic. Of the N, r say he is a schizophrenic. If two psychiatrists are chosen at random from the N, what is the probability that
    (a) They both say the patient is a schizophrenic?
    (b) That one says he is a schizophrenic and the other does not?

20. Consider a set of eight cards consisting of the four jacks and the four queens from a standard bridge deck. Two cards are drawn at random (WOR).
    (a) Describe an appropriate sample space.
    (b) Find the probability that
        (i) Both cards are jacks given that at least one is a jack.
        (ii) Both cards are jacks given that exactly one is a red jack.
        (iii) Both cards are jacks given that one is the jack of hearts.

21. According to a lifetable of 100,000 persons living at age 10, 95,000 live to be 16 years of age and 69,800 live to be 50 years of age. If you are 16 years of age, what are your chances of living to be 50 years of age?

22. In a group of 50 language students, it is known that 30 read French, 20 read German, five read Russian, two read French and Russian, three read German and Russian, two read German and French and one reads all three languages. If a student is selected at random, what is the probability that he studies German if

22. (a) He studies French?
    (b) He studies French and Russian?
    (c) He studies neither French nor Russian?

23. A student is given two sets (A and B) of questions to study in preparation for a test. Set A consists of five questions and set B has 10 questions. The student knows the answer to three questions in set A and nine in set B. The test will consist of three questions chosen by the instructor.
    (a) Suppose the instructor randomly selects a set of questions and then chooses three questions at random. If the instructor is twice as likely to choose set B as set A, what is the probability that the student can answer all three questions?
    (b) Suppose instead the instructor chooses three questions at random from the combined list of 15 questions. What is the probability that the student can answer all three questions?
    (c) If you were the student, which selection scheme, (a) or (b), would you prefer? Why?

24. (a) A lot of 25 items is to be inspected by means of a two-stage sampling plan. A sample of five items is drawn. If one or more is bad, the lot is rejected. If all are nondefective, a second sample of 10 items is drawn from the remaining 20 items. The lot is then rejected if any item in the second sample is bad; otherwise it is accepted. Find the probability of accepting a lot which has two defective items.
    (b) An alternative single-stage plan for this problem might be to choose a single sample of 15 items and accept the lot only if there are no defectives in the sample. Find the probability of accepting a lot of 25 with two defectives and compare this probability with that found in (a).

25. A box contains five red and three green balls. One ball is drawn at random. It is replaced and two more balls of this same color are added to the box. A second ball is then drawn.
    (a) Set up a two-stage model for this experiment and discuss the probability models (outcomes and probabilities) corresponding to each stage.
    (b) Find the probability of a red ball on the second draw.
    (c) Find the probability of one ball of each color in the two draws.
    (d) If the second ball drawn is red, find the probability that the first ball drawn was red.

26. As a generalization of Problem 25, suppose the box had r red balls and g green ones. A ball is drawn at random. This ball is returned to the box and c balls of the same color are added. A second draw is then made.
    (a) Construct an appropriate two-stage model.
    (b) Show that Pr(red ball on second draw | red ball on first draw) is greater than Pr(red ball on first draw).
    (c) Show that the probability of a red ball on the second draw is the same as that for a red ball on the first draw.

27. Prove that
    (a) $\Pr[(A \cup B) \mid C] = \Pr(A \mid C) + \Pr(B \mid C) - \Pr[(A \cap B) \mid C]$
    (b) $\Pr[(\bar{A} \cup \bar{B} \mid C] = 1 - \Pr[(A \cap B) \mid C]$.

28. Reread Problem 17 in Chapter 2. Find the probability that a person selected at random <u>from those who read newspaper C</u>
    (a) reads newspaper A or B
    (b) reads neither newspaper A nor B.

29. A firm which has seven branches in widely separated locations wishes to interview a sample of its workers. They decide to interview workers from only three branches. From each of the chosen branches a random sample of 10% of the workers will be chosen. Set up an appropriate two-stage model to describe this sampling procedure.

30. A box contains three defective and seven nondefective light bulbs.
    (a) If three bulbs are drawn consecutively at random (WOR), develop a three-stage model (outcomes and probabilities). You may find a tree diagram useful.
    (b) Determine the probability that
        (i) At least one nondefective bulb is drawn.
        (ii) All three are nondefective if the first is nondefective.

31. The probability that Jack wins at ping-pong is a function of his previous performance. The probability that he wins a given game is $(1/2)^{k+1}$, where k is the number of games he has won so far. What is the probability that he will win at least two if he plays three games?

32. During the month of May the probability of a rainy day is .2. Suppose the Expos win on a clear day with probability .7 and on a rainy day with probability .4. If we know that they won on a certain day in May, what is the probability that it was a rainy day?

33. A businessman receives a postdated cheque in payment of a bill. From experience he has learned that even people with sufficient funds to cover a cheque accidentally postdate it. He estimates that this occurs about twice in 100 times. Those customers with insufficient funds invariably postdate their cheques. The businessman estimates that about 90% of his customers have sufficient funds. What is the probability that the postdated cheque is from a customer with sufficient funds?

34. A fruit store obtains its supply of apples from three producers, X, Y and Z. Producer X supplies 20% of the requirements and on the average 2% of his applies are bruised. Producers Y and Z supply 30 and 50%, respectively. On the average 1% of Y's and 5% of Z's apples are bruised. In a given week the fruit store owner purchases 5000 apples. If an apple chosen at

random turns out to be bruised, what is the probability that it was supplied by Y?

35. A rat is allowed to choose one of four mazes at random. If we know that the probabilities of his getting through the various mazes in two minutes are .5, .3, .2, .1 and we find that the rat has escaped after two minutes, how likely is it that he chose the first maze? The second maze?

36. Two closely related species of mushrooms (I and II) are difficult to identify without the aid of a microscope. One method used in the field to separate the two species is to note the presence or absence of a ring on the stalk of the plant. Ninety percent of species I and 20% of species II have the ring. It is also known that in the particular area where the mushrooms are being studied, 70% of them are species I.
    (a) Suppose the field worker finds a mushroom with a ring and decides it belongs to species I. What is the probability he is correct?
    (b) If mushrooms with rings are classified as species I and those without as species II, what proportion of mushrooms will be correctly classified?

37. A diagnostic procedure detects 95% of those people who in fact have the disease. It gives a "false-positive" in 10% of the people who do not have the disease. It is known that the overall incidence of the disease is 1%.
    (a) If a patient is chosen at random and the test gives a positive result, what is the probability he actually has the disease?
    (b) With this procedure, describe what is meant by making an error?
    (c) Calculate the probability of making an "error" with this testing procedure.

38. Repeat Problem 37 assuming that the incidence of the disease in the population is 50%. Compare your results with those obtained in the previous problem.

Chapter 5

Independence

5-1 Independence of Two Events

In everyday language we describe two events that "have nothing to do with one another" as *independent events*. In order to develop a more statistical definitions for independence, let us consider the following examples.

Example 1: Of six trees planted in a straight line, we know that two are diseased. (a) If any one of the trees is equally likely to be diseased, what is the probability that the diseased trees are side by side? (b) If we know that tree 3 is one of the diseased ones, what is the probability that the diseased trees are adjacent? (c) Now if the trees are planted in a circle with 1 and 6 adjacent and 3 is known diseased, what is the probability that the diseased trees are adjacent?

Solution: (a) For ease of solution, suppose the trees are numbered 1 to 6 as they stand. The sample space S consists of all the subsets of size 2 that can be formed:

$$S = \{(1\ \ 2), (1\ \ 3), (1\ \ 4), (1\ \ 5), (1\ \ 6), (2\ \ 3), (2\ \ 4), (2\ \ 5), (2\ \ 6), (3\ \ 4), (3\ \ 5), (3\ \ 6), (4\ \ 5), (4\ \ 6), (5\ \ 6)\}.$$

If S represents all the possible pairs of diseased trees, five of these pairs contain adjacent trees: (1 2), (2 3), (3 4), (4 5), (5 6). Let A denote the event that the diseased trees are adjacent.

Then $Pr(A) = 5/15 = 1/3$. (b) Now condition on the event B: tree 3 is diseased. Recall that

$$Pr(A \mid B) = \frac{Pr(A \cap B)}{Pr(B)} .$$

Referring to S we see that the event B contains the five points (1 3), (2 3), (3 4), (3 5), (3 6). The joint event $(A \cap B)$ occurs whenever tree 3 and a tree adjacent to it are diseased; namely, (2 3), (3 4). Combining these results, we have

$$Pr(A \mid B) = \frac{2/15}{5/15} = \frac{2}{5} .$$

We have solved this problem by considering the definition of conditional probability. Can you solve it by constructing a reduced sample space? (c) How does planting the trees in a circle affect the solution? The sample space remains the same. Since the event A now has an extra point (1 6),

$$Pr(A) = \frac{2}{5}$$

$$Pr(A \mid B) = \frac{2/15}{5/15} = \frac{2}{5} .$$

This demonstrates that when the trees are planted in a circle, $Pr(A)$ and $Pr(A \mid B)$ are equal.

When the trees were planted in a straight line, the fact that tree 3 was known to be diseased changed the probability of adjacent trees being diseased:

$$Pr(A \mid B) \neq Pr(A).$$

In the case of circular planting, the fact that we know tree 3 was diseased did not provide any further information:

$$Pr(A \mid B) = Pr(A).$$

Here we would say that the events A and B are *independent* or unrelated.

Example 2: Suppose we toss two balanced coins (a penny and a nickel). Let the events E and F be defined as

E: head on penny

F: head on one coin and tail on the other

Are the events E and F independent?

Solution: If we refer to Section 2-1, we will recall that an appropriate sample space is

$$S = \{H_P H_N, H_P T_N, T_P H_N, T_P T_N\}$$

with each outcome having equal probability. If E and F are independent, then $Pr(E \mid F)$ and $Pr(E)$ should be equal. Clearly, by examination of S,

$$Pr(E) = Pr(H_P H_N \text{ or } H_P T_N) = \frac{1}{2} .$$

If F has occurred, the sample space is reduced to

$$S^* = \{H_P T_N, T_P H_N\}$$

with each outcome equally likely. Then

$$Pr(E \mid F) = \frac{1}{2} .$$

Using these examples as a model, we will define independent events in a statistical sense.

Definition 5·1: *Two events* E *and* F *in a sample space* S *are (statistically) independent if*

$$Pr(E \mid F) = Pr(E).$$

Since $Pr(E \cap F)$ can be written as $Pr(E \mid F)Pr(F)$ or $Pr(F \mid E)Pr(E)$, Definition 5·1 could have alternatively been stated

$$Pr(F \mid E) = Pr(F).$$

In the following discussion of independent events we will usually drop the term statistically. We will, however, be using the term independence in its technical sense.

A special form of the multiplication rule follows from Definition 5·1.

> Multiplication Rule for Independent Events: *If two events* E *and* F *are independent, then*
>
> $$\Pr(E \cap F) = \Pr(E)\Pr(F).$$

Proof:

In general we know that

$$\Pr(E \cap F) = \Pr(E \mid F)\Pr(F).$$

If, however, E and F are independent, then

$$\Pr(E \mid F) = \Pr(E).$$

Substituting for $\Pr(E \mid F)$ in terms of the unconditional probability, we have

$$\Pr(E \cap F) = \Pr(E)\Pr(F). \qquad \square$$

The converse of this law is also true: If $\Pr(E \cap F) = \Pr(E)\Pr(F)$, *then* E *and* F *are independent*. Its proof is left as an exercise for the student.

In summary, independence can be established in two possible ways:

1. Showing that the conditional probability $\Pr(E \mid F)$ is equal to the unconditional probability $\Pr(E)$ or $\Pr(F \mid E) = \Pr(F)$.
2. Demonstrating that the joint probability $\Pr(E \cap F)$ is equal to the product of the unconditional probabilities, $\Pr(E)$ and $\Pr(F)$.

Now suppose that we know that the events E and F are independent. What can we say about the independence of the pairs of events: E and $\bar{F}$, $\bar{E}$ and F, $\bar{E}$ and $\bar{F}$?

> Theorem 5·1: *If* E *and* F *are two independent events in* S, *then the pairs of events* E *and* $\bar{F}$, $\bar{E}$ *and* F, $\bar{E}$ *and* $\bar{F}$ *are each independent.*

Proof:

The probabilities of these events can be summarized in the following 2 × 2 table:

	F	$\bar{F}$	
E	Pr(E)Pr(F)	?	Pr(E)
$\bar{E}$	?	?	Pr($\bar{E}$)
	Pr(F)	Pr($\bar{F}$)	

Since E and F are independent, the entry $\Pr(E \cap F)$ can be written as Pr(E)Pr(F). How can we find $\Pr(E \cap \bar{F})$? Recall that the event E and its probability can be written as

$$E = (E \cap F) \cup (E \cap \bar{F}).$$

Since the events on the r.h.s. are m.e., the

$$\Pr(E) = \Pr(E \cap F) + \Pr(E \cap \bar{F}).$$

Writing $\Pr(E \cap F)$ as the product of the individual probabilities and solving for $\Pr(E \cap \bar{F})$, we get

$$\Pr(E \cap \bar{F}) = \Pr(E) - \Pr(E)\Pr(F) = \Pr(E)[1 - \Pr(F)].$$

Since $\Pr(\bar{F})$ by definition is $1 - \Pr(F)$, we can write

$$\Pr(E \cap \bar{F}) = \Pr(E)\Pr(\bar{F}).$$

Therefore, E and $\bar{F}$ are independent. Similarly, we have

$$\begin{aligned}\Pr(\bar{E} \cap F) &= \Pr(F) - \Pr(E \cap F)\\ &= \Pr(F) - \Pr(E)\Pr(F)\\ &= \Pr(F)[1 - \Pr(E)] = \Pr(F)\Pr(\bar{E}),\end{aligned}$$

which shows that $\bar{E}$ and F are independent. Likewise, $\bar{E}$ and $\bar{F}$ are independent since

$$\begin{aligned} Pr(\bar{E} \cap \bar{F}) &= Pr(\bar{E}) - Pr(\bar{E} \cap F) \\ &= Pr(\bar{E}) - Pr(\bar{E})Pr(F) \\ &= Pr(\bar{E})[1 - Pr(F)] = Pr(\bar{E})Pr(\bar{F}). \quad \square \end{aligned}$$

The student should be able to justify the steps on the r.h.s. of the above equations.

Theorem 5·1 demonstrates that if the events E and F are independent, then the probabilities for the four intersection events in a 2 × 2 table can be found by taking the product of the corresponding marginal probabilities.

5-2 Independence For More Than Two Events

Suppose we are interested in m events $A_1, A_2, \ldots, A_m$. A natural generalization of independence to more than two events would be: $A_1, A_2, \ldots, A_m$ are independent if

$$Pr(A_1 \cap A_2 \cap \cdots \cap A_m) = Pr(A_1)Pr(A_2)\cdots Pr(A_m).$$

We shall see, however, that if $m \geq 3$, this condition alone is not sufficient to guarantee the truth of some of the probability laws we have developed for two events. For more than two events, we have

> Definition 5·2: *The* m *events* $A_1, A_2, \ldots, A_m$ *are said to be* <u>mutually independent</u> *if the joint probability of every combination of the events taken any number at a time is equal to the product of their individual probabilities.*

Let us restrict ourselves to m = 3 events: A_1, A_2, A_3. Mutual independence means

$$\Pr(A_1 \cap A_2 \cap A_3) = \Pr(A_1)\Pr(A_2)\Pr(A_3)$$

$$\Pr(A_1 \cap A_2) = \Pr(A_1)\Pr(A_2)$$

$$\Pr(A_2 \cap A_3) = \Pr(A_2)\Pr(A_3)$$

$$\Pr(A_1 \cap A_3) = \Pr(A_1)\Pr(A_3).$$

All four of these conditions must hold. In the case in which only the last three conditions hold, the events A_1, A_2 and A_3 are said to be pairwise independent. Example 3 illustrates that pairwise independence does not imply mutual independence. In Example 4 we see that, although the first condition holds, the events are not pairwise independent.

Example 3: Let us return to the tossing of a penny and nickel and define the events:

A_1: head on penny

A_2: head on nickel

A_3: coins match

Are the events A_1, A_2 and A_3 mutually independent?

Solution: From the sample space and its probabilities, we see that

$$\Pr(A_1) = \Pr(A_2) = \Pr(A_3) = \frac{1}{2},$$

and

$$\Pr(A_1 \cap A_2) = \Pr(A_1 \cap A_3) = \Pr(A_2 \cap A_3) = \frac{1}{4}.$$

Thus these events are pairwise independent; however,

$$\Pr(A_1 \cap A_2 \cap A_3) = \frac{1}{4} \neq \Pr(A_1)\Pr(A_2)\Pr(A_3).$$

Although the events are pairwise independent, they are not mutually independent.

Example 4: In the town of Uban three editions of the daily paper are published: morning (M), evening (E) and weekend (W). Suppose that the probability associated with a person's subscribing to the papers can be illustrated by the following Venn diagram. Are the events M, E and W mutually independent?

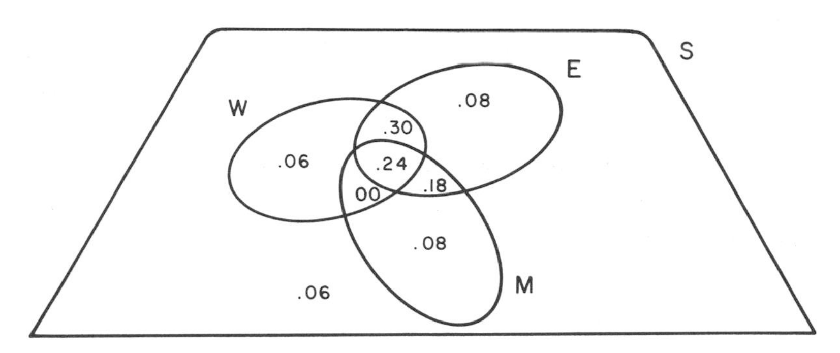

Solution: From the above diagram, we can easily determine the probabilities: $\Pr(W) = .6$, $\Pr(E) = .8$, $\Pr(M) = .50$, $\Pr(W \cap E) = .54$, $\Pr(W \cap M) = .24$, $\Pr(E \cap M) = .42$, $\Pr(W \cap E \cap M) = .24$. Clearly the events are <u>not</u> pairwise independent, but the

$$\Pr(W \cap E \cap M) = \Pr(W)\Pr(E)\Pr(M).$$

Again, these events are not mutually independent.

5-3 Probabilities Associated with Mutually Independent Events

In statistical applications we are usually interested in situations in which the events are mutually independent. In the following discussion we shall refer to such events as being *independent*. At all times *mutually independent* will be implied.

Under the assumption of independence, we can easily extend the multiplication law for two independent events.

Multiplication Law for m Independent Events: *If* $A_1, A_2, \ldots, A_m$ *are* m *independent events, then*

$$\Pr(A_1 \cap A_2 \cap \cdots \cap A_m) = \Pr(A_1)\Pr(A_2)\cdots\Pr(A_m).$$

Suppose an experiment consists of a repetition of independent trials. Then to find the probability of joint events we can apply the above law.

Example 5: A box contains r red marbles and b black marbles. If a game consists of drawing with replacement until a red marble appears, what is the probability of winning on the kth draw?

Solution: Since the draws are WR, we can visualize the experiment as consisting of k independent trials. For each trial, Pr(red) = r/(r + b) and Pr(black) = b/(r + b). Winning on the kth draw results from the first (k - 1) draws being black marbles and the kth being a red marble. Hence,

$$\Pr(\text{winning on kth draw}) = \left(\frac{b}{r+b}\right)^{k-1}\left(\frac{r}{r+b}\right).$$

Example 6: Suppose a balanced die is rolled three times. (a) What is the probability that we get a 4, 1 and 1 in this particular order? (b) What is the probability we get a 4, 1 and 1 in any order?

Solution: (a) This problem is similar to Example 5. The three tosses are three independent trials; hence,

$$\Pr(4, 1, 1 \text{ in that order}) = \left(\frac{1}{6}\right)^3.$$

(b) How is part (b) different from (a)? Here we are interested in all the possible orders in which a 4 and two 1's can arise. Since a 4 and two 1's can be reordered in 3!/2! ways,

$$\Pr(4, 1, 1 \text{ in any order}) = 3\left(\frac{1}{6}\right)^3.$$

Example 7: A biased die has been mixed with two other balanced ones. It is known that the unbalanced one rolls 6 twice as often as any other outcome. If one of the three dice is selected at random and rolled five times, what is the probability (a) that the die picked is the biased one if <u>all</u> five rolls are 6's? (b) that the die chosen is the biased one if <u>none</u> of the five rolls is a 6?

Solution: This example combines independence and Bayes' Rule. The prior probability of a biased and unbiased die is given by

$$\Pr(\text{biased}) = \frac{1}{3}$$

$$\Pr(\text{unbiased}) = \frac{2}{3}\ .$$

(a) Let us define the event

$$A = \text{five 6's in five rolls.}$$

Then

$$\Pr(A) = \Pr(A \cap \text{unbiased}) + \Pr(A \cap \text{biased})$$
$$= \frac{2}{3}\left(\frac{1}{6}\right)^5 + \frac{1}{3}\left(\frac{2}{7}\right)^5.$$

Using Bayes' Rule

$$\Pr(\text{biased} \mid A) = \frac{\frac{1}{3}\left(\frac{2}{7}\right)^5}{\frac{2}{3}\left(\frac{1}{6}\right)^5 + \frac{1}{3}\left(\frac{2}{7}\right)^5}\ .$$

(b) In this case let B be the event

$$B = \text{no 6 in five rolls}$$

with

$$\Pr(B) = \frac{2}{3}\left(\frac{5}{6}\right)^5 + \frac{1}{3}\left(\frac{5}{7}\right)^5.$$

The posterior probability

$$\Pr(\text{biased} \mid B) = \frac{\frac{1}{3}\left(\frac{5}{7}\right)^5}{\frac{2}{3}\left(\frac{5}{6}\right)^5 + \frac{1}{3}\left(\frac{5}{7}\right)^5}\ .$$

Which event, A or B, do you feel is more useful an observation if you are trying to modify the prior probabilities?

Example 8: In electrical systems we are often interested in tracing current from one point, through several component parts, to another point. The system can be constructed either in series or in parallel:

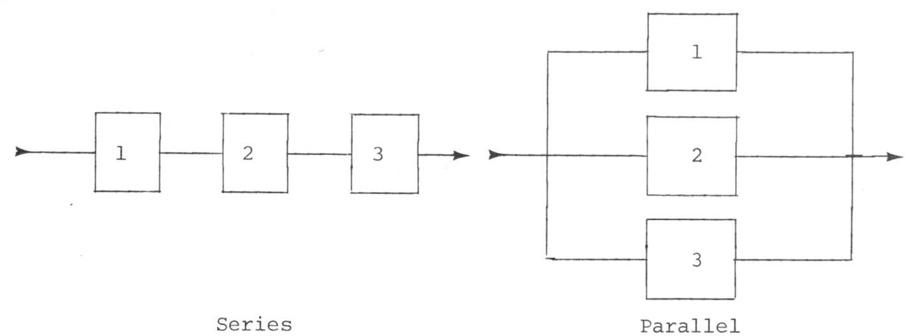

In the series system the current must be able to flow through each component; hence, failure of at least one component causes the system to fail. If, however, the components are arranged in parallel, the current may flow through any one of the components and the system fails only when all the components fail.

Consider a system with k components and let A_i represent the failure of the ith component. Failure of the system can be expressed as

$$A_1 \cup A_2 \cup \cdots \cup A_k$$

when the system is arranged in a series and

$$A_1 \cap A_2 \cap \cdots \cap A_k$$

when it is arranged in parallel.

Often the term <u>reliability</u> is used to describe the probability that the system works:

$$\begin{aligned}\text{Reliability (series system)} &= \Pr\{\overline{(A_1 \cup A_2 \cup \cdots \cup A_k)}\}\\ &= 1 - \Pr\{(A_1 \cup A_2 \cup \cdots \cup A_k)\}\\ &= \Pr\{(\bar{A}_1 \cap \bar{A}_2 \cap \cdots \cap \bar{A}_k)\},\end{aligned}$$

the probability that all of the components work.

$$\text{Reliability (parallel system)} = \Pr\{\overline{(A_1 \cap A_2 \cap \cdots \cap A_k)}\}$$
$$= 1 - \Pr\{(A_1 \cap A_2 \cap \cdots \cap A_k)\}$$
$$= \Pr\{(\bar{A}_1 \cup \bar{A}_2 \cup \cdots \cup \bar{A}_k)\},$$

the probability that at least one component works.

Now in the special case when the k components operate <u>independently</u>,

$$\text{Reliability (series)} = \Pr(\bar{A}_1)\Pr(\bar{A}_2) \cdots \Pr(\bar{A}_k)$$
$$\text{Reliability (parallel)} = 1 - \{\Pr(A_1)\Pr(A_2) \cdots \Pr(A_k)\}.$$

To illustrate these concepts consider two parallel systems arranged in a series:

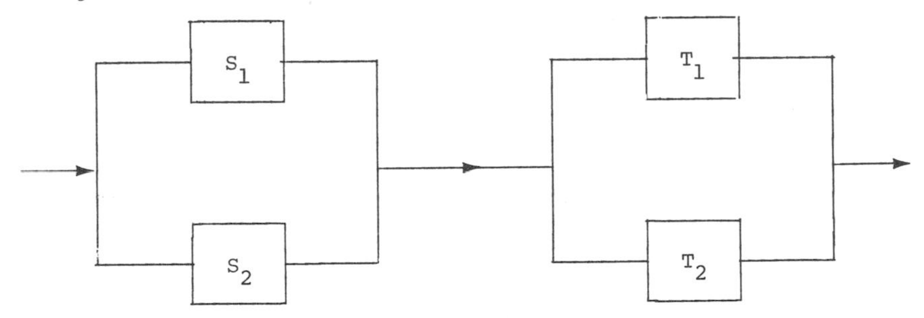

Suppose the probability that a type S component functions is p_S, while the corresponding probability for a type T component is p_T. The four components S_1, S_2, T_1, T_2 operate independently of one another. What is the reliability of the system?

Let S_1, S_2, T_1, T_2 represent the events that the four components <u>fail</u>. Then symbolically the system will <u>fail</u> if

$$(S_1 \cap S_2) \cup (T_1 \cap T_2).$$

The reliability of the system is the probability of the complement of this event:

$$\Pr\{\overline{(S_1 \cap S_2) \cup (T_1 \cap T_2)}\}.$$

Using DeMorgan's laws (see Appendix B) and the mutual independence of these events, we get

$$\begin{aligned}\text{Reliability} &= \Pr[\overline{(S_1 \cap S_2)} \cap \overline{(T_1 \cap T_2)}] \\ &= \Pr(\overline{S_1 \cap S_2})\Pr(\overline{T_1 \cap T_2}) \\ &= \Pr(\bar{S}_1 \cup \bar{S}_2)\Pr(\bar{T}_1 \cup \bar{T}_2) \\ &= (2p_S - p_S^2)(2p_T - p_T^2).\end{aligned}$$

Example 9: The concepts of probability have come to play an increasingly important role in formulating models for random phenomena in the biological, physical and social sciences. For this example the ideas of conditional probability and independence will be applied to a problem in genetics.

Based on experiments using peas, Gregor Mendel first pointed out that inheritance of certain traits could be regarded as a random experiment. He postulated that biological characteristics are transmitted to progeny by entities called genes. These genes appear in pairs with one coming from each parent.

In the simplest model each gene of a pair can assume only one of two forms: recessive (a) or dominant (A). Assuming then that a characteristic is determined by a pair of these genes, three distinct combinations or genotypes are possible: (1) AA, the pure dominant; (2) Aa, the hybrid; (3) aa, the pure recessive. Although three distinct genotypes are possible, it is often the case that the AA and Aa individuals are not distinguishable from one another in physical appearance. Geneticists say that AA and Aa individuals have the same phenotype.

We know that offspring receive one gene from each parent. What type of progeny will result if individuals with these different genotypes mate?

AA × AA → AA

aa × aa → aa

AA × Aa → AA or Aa

Aa × Aa → AA, Aa or aa

What probabilities should we attach to each of these outcomes? If we assume that (1) the offspring of an Aa parent has an equal probability of receiving either A or a and (2) genes transmitted by the two parents are unrelated, then the probability model for the AA × Aa and Aa × Aa matings are

AA × Aa:

Outcome (E_i)	AA	Aa
$\Pr(E_i)$	$\frac{1}{2}$	$\frac{1}{2}$

Aa × Aa:

Outcome (E_i)	AA	Aa	aa
$\Pr(E_i)$	$\frac{1}{4}$	$\frac{1}{2}$	$\frac{1}{4}$

Suppose the recessive characteristic a is less viable than A, such that individuals with genotype Aa will produce genes a and A with probabilities:

$$\Pr(a) = \frac{\alpha}{2} \quad \text{and} \quad \Pr(A) = 1 - \frac{\alpha}{2},$$

with $0 < \alpha < 1$. A cross of the type Aa × Aa then has the probability model

Outcome (E_i)	AA	Aa	aa
$\Pr(E_i)$	$\frac{1}{4}(2-\alpha)^2$	$\frac{1}{2}\alpha(2-\alpha)$	$\frac{1}{4}\alpha^2$

Verification of these results is left as an exercise.

Problems

1. Show that if $\Pr(E \cap F) = \Pr(E)\Pr(F)$, then E and F must be independent.
2. A community has two police cars, which operate independently of one another. The probability that a specific car will be available when needed is .99.

(a) What is the probability that neither car is available when needed?

(b) What is the probability that a car is available when needed?

3. Would you expect the following events to be independent? Why or why not?

   (a) Outcomes of one throw each of two different dice.

   (b) Outcomes of two successive throws of the same die.

   (c) Outcomes of drawing two balls with replacement from a box of r red and g green balls.

   (d) Outcomes of drawing two balls without replacement from a box of r red and g green balls.

4. Are the events G and H discussed in Problem 10 of Chapter 4 independent?

5. Show that if A and B are mutually exclusive, they cannot be independent unless Pr(A) = 0 or Pr(B) = 0.

6. Consider the following 2 × 2 table:

	Smokers	Nonsmokers
Males	4	6
Females	6	x

   How many individuals must be nonsmoking females if the characteristics smoking and sex are to be independent?

7. Two red tulip bulbs are accidentally mixed with six white ones. In the autumn the eight bulbs are planted in two rows of four bulbs each.

   (a) What is the probability that the two red tulips are planted on diagonal corners?

   (b) In the spring if the first tulip to bloom is red and it appears in a corner, what is the probability that the other red tulip will appear in the diagonal corner?

8. Suppose a classroom has nine desks arranged in three rows of three desks each. The instructor chooses two students at random. Let A be the event that both students chosen sit in corner desks and B be the event that both students chosen are from the same row.
   (a) Find Pr(A).
   (b) Find Pr(B).
   (c) Are the events A and B independent?

9. If A and B are mutually exclusive, show that

$$\Pr(A \mid \bar{B}) = \frac{\Pr(A)}{1 - \Pr(B)} .$$

   Would this necessarily be true if A and B are not mutually exclusive?

10. A political scientist has undertaken a research project to investigate the voting patterns of husbands and wives. Suppose that the probability that a married man votes is .45 while the probability that a married woman votes is .40. In contrast, the probability that a married woman votes given that her husband votes is .60.
   (a) Find the probability both husband and wife vote.
   (b) Are the actions of the husband and wife independent?
   (c) Find the probability that the husband votes given that his wife has voted. Comment on this result.

11. In a tennis match it is assumed that a player's performance in an individual game is unrelated to previous performance. Suppose a balanced coin is tossed to decide which player A or B begins by serving. The players then alternate in serving. If the probability that A wins in a game in which he is serving is .8 and the probability that he loses in a game in which he is receiving is .6, what is the probability that A will win the first six games?

12. To test the effectiveness of a new headache remedy, a large sample of n people was selected. Since it was thought that a person's reaction might be sex-related, the data collected were sex-specific. The following table summarizes the results:

		Drug effect	
		E	$\bar{E}$
Sex	M	a	b
	F	c	d

with a + b + c + d = n.

(a) Based on these data, determine

(i) Pr(M), (ii) Pr(E), (iii) Pr(M | E),

(iv) Pr(E | M).

(b) When will the events E and M be independent? Show that if ad - bc = 0, E and M will be independent.

(c) If E and M are independent, we say that the drug effect and sex factor are independent. Why can we make this statement?

13. A study is undertaken to demonstrate the effectiveness of job training. Two skilled workers and two novices are each assigned the same complicated task. Assuming that the workers operate independently and that the chance that a skilled worker successfully completes the task is .8, whereas for the novice the chance is .5, find the probability that

(a) All four complete the task.

(b) Two complete the task.

(c) The number of skilled workers who complete the task equals the number of novices who complete the task.

14. In Problem 13 suppose that training has no effect; that is, the skilled and the unskilled worker each has a 50% chance of completing the task. Under this assumption answer questions (a) through (c). Comment on the role of job training in light of these results.

15. If the events E and F are independent, determine

$$\Pr[(E \cap F) \mid F].$$

How would you interpret this result?

16. From past experience it is known that out of every 10 undergraduate university textbooks published only one will be an outstanding success. A publisher has chosen six books for publication.

    (a) What is the probability that
        (i) All will be outstanding successes?
        (ii) None will be an outstanding success?
        (iii) Exactly one will be an outstanding success?
        (iv) At least one will be an outstanding success?
        (v) At least two will be outstanding successes?

    (b) What basic assumption have you made in answering part (a)?

17. Generalize Problem 16 in the following way: Let the probability of an outstanding success be p. If n books are selected, determine the probability that

    (a) All are outstanding successes.
    (b) None is an outstanding success.
    (c) Exactly one is an outstanding success.
    (d) At least one is an outstanding success.
    (e) Exactly k are outstanding successes.

18. At a carnival a prize is offered to any person who in a sequence of three independent trials hits the target twice in a row.

    (a) What is the probability that Jack wins a prize if his probability of hitting on any shot is p?
    (b) Now suppose the probabilities vary with each trial; that is, p_1, p_2 and p_3. Find the probability that Jack wins a prize.

19. A box contains three red and five white marbles. Two marbles are sequentially selected. Consider the event

R_1: Red on first draw

R_2: Red on second draw.

Determine whether or not R_1 and R_2 are independent when the selection is

(a) With replacement.

(b) Without replacement.

20. A chain consists of k links. The strengths of the links are mutually independent. If the probability that any one link fails under a specified load is θ, what is the probability that the chain will fail under the load?

21. Explain carefully what is meant by the mutual independence of any three events A, B and C.

22. Construct an example in which each of the probabilities of the three events A, B and C is positive and the events are pairwise independent but $\Pr(A \cap B \cap C) = 0$. Draw a Venn diagram illustrating these results.

23. Consider the following table of events and their corresponding probabilities:

Event	$A \cap B \cap C$	$A \cap \bar{B} \cap C$	$A \cap B \cap \bar{C}$	$A \cap \bar{B} \cap \bar{C}$
Probability	1/16	5/16	3/16	1/8

Event	$\bar{A} \cap B \cap C$	$\bar{A} \cap B \cap \bar{C}$	$\bar{A} \cap \bar{B} \cap C$	$\bar{A} \cap \bar{B} \cap \bar{C}$
Probability	1/8	1/16	1/16	1/16

(a) Find Pr(A), Pr(B) and Pr(C).

(b) Are A, B and C mutually independent?

(c) Find $\Pr(A \mid B)$. Are A and B independent?

(d) If Pr(A) = 1/2, Pr(B) = 1/3 and Pr(C) = 1/6, what must be the entries in the above table in order for A, B and C to be mutually independent?

24. If the events A, B and C are pairwise independent and the event $(A \cap B)$ and C are independent, show that the events A, B and C are mutually independent?

25. Suppose box I contains two red balls and three white ones while box II contains one red and four white ones. A box is chosen at random by selecting a random number from 0 through 9. If a 1 or 2 is selected, box I is chosen; otherwise box II is chosen. If a ball is then selected at random from the box and found to be red, what is the probability that it came from box I? From box II?

26. Reconsider Problem 25 with the alteration that now n balls were chosen at random with replacement after having selected box I or II as before.
    (a) If all n balls are red, what is the posterior probability associated with each box?
    (b) For what value or values of n will these posterior probabilities be equal?
    (c) As n becomes large, what happens to these probabilities? What is your interpretation of these results?

27. Review Example 8 and label the system described there as system I. Now consider system II, which is two series systems arranged in parallel:

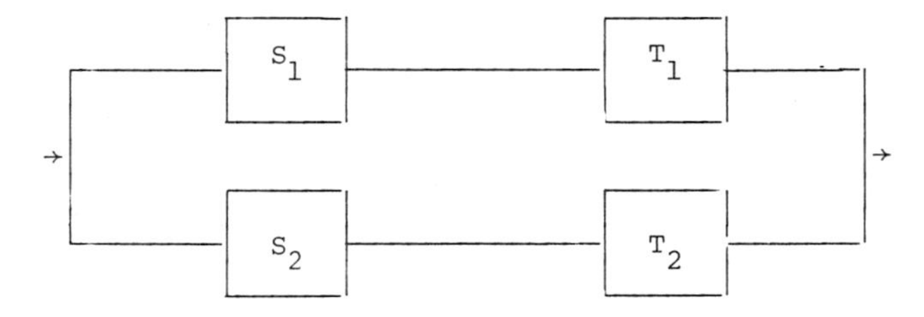

    (a) Find the reliability of system II using the assumptions of Example 8.
    (b) Under what conditions will system I be more reliable than system II?

28. Explain in detail how the probabilities given in the last table of Example 9 were obtained.

29. In an experiment, characteristic A is dominant while a is recessive. Consider two types of mating:

$$\text{I:} \quad AA \times aa$$
$$\text{II:} \quad Aa \times Aa.$$

Suppose a large number n of each type of mating are mixed together.

(a) What is the probability that an individual selected at random is Aa?

(b) Given that the offspring is Aa, what is the probability that he is progeny of type I mating?

30. Extend Problem 29 to the situation in which 1/3 of the offspring pool is from type I matings and the remaining 2/3 is from type II matings.

Chapter 6

Random Variables

6-1 Quantifying the Random Experiment

In describing a random experiment we have been using as much detail as possible in constructing the sample space and associated probabilities. Frequently, however, we may not be interested in all the details of the experiment. Our main concern may be to describe the experiment by some numerical quantity rather than in great detail.

When, for example, we are deciding on the quality of a "lot" of mass-produced items by examining a random sample, we are interested only in the number of defectives in the sample. The condition of a particular item (defective or nondefective) is not of much interest since we are actually concerned with making decisions or drawing inferences about the whole population of items produced.

In a situation such as this we are quantifying, or assigning a numerical value to, the outcome of the random experiment. We shall refer to this numerical description of the random experiment as a *random variable*.

Just as each outcome in the sample space had a probability associated with it, the random variable will take on a particular numerical value with some associated probability. Analogous to the concept of a probability model which we discussed in connection with the sample space, here we will combine the values of the random variable and its probabilities by means of the *probability function*.

To further illustrate the concepts of random variable (r.v.) and probability function (p.f.), let us consider the sex distribution of three-children families. We know that a possible sample space is the set of ordered triplets:

$$S = \{(M, M, M), (M, M, F), (M, F, M), (F, M, M), (F, M, F), (F, F, M), (M, F, F), (F, F, F)\}.$$

Assuming the uniform model or that the births are independent with Pr(M) = Pr(F) = 1/2 for each birth, we have each point in S equally likely with probability 1/8.

Now suppose we wish to quantify the results. Let us define a random variable:

$$X = \text{number of males.}$$

Here X can take on the values 0, 1, 2 or 3. We will speak of the set {0, 1, 2, 3} as the value set of X. The probabilities associated with each of these values can be found by considering S.

Outcome in S: E_i	$Pr(E_i)$	x	Pr(X = x)
(M, M, M)	$\frac{1}{8}$	3	$\frac{1}{8}$
(M, M, F)	$\frac{1}{8}$		
(M, F, M)	$\frac{1}{8}$	2	$\frac{3}{8}$
(F, M, M)	$\frac{1}{8}$		
(F, F, M)	$\frac{1}{8}$		
(F, M, F)	$\frac{1}{8}$	1	$\frac{3}{8}$
(M, F, F)	$\frac{1}{8}$		
(F, F, F)	$\frac{1}{8}$	0	$\frac{1}{8}$

At first it may seem somewhat awkward to distinguish between the random variable and the values it assumes. To help eliminate this confusion, we will use the conventional notation that the capital letter, say X, represents the random variable and the lower case letter x designates one of its values. We will adopt the functional notation f(x) (read "f of x") for the probability that the random variable X takes on the value x:

$$f(x) = \Pr(X = x).$$

Thus the probability function of X in our example can be summarized as

1. Tabular form:

x	0	1	2	3
f(x)	$\frac{1}{8}$	$\frac{3}{8}$	$\frac{3}{8}$	$\frac{1}{8}$

2. Graphical form:

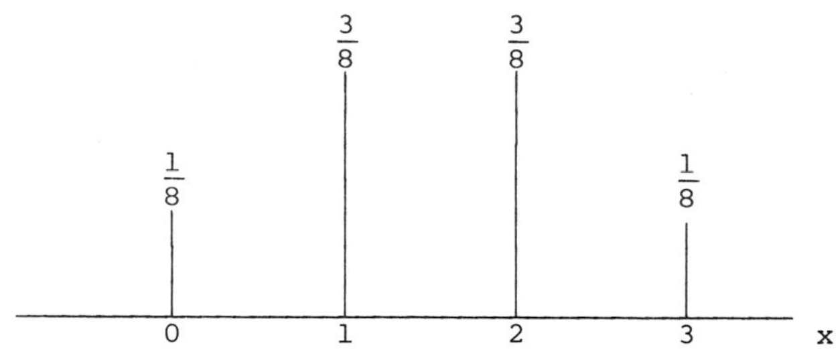

3. Mathematical formula:

$$f(x) = \binom{3}{x}\left(\frac{1}{2}\right)^3 \qquad x = 0, 1, 2, 3.$$

By substituting each value of x, we see that this formula generates the same values for f(x) as those in the tabular and graphical forms. We can now formally define the terms random variables and probability function.

> Definition 6·1: *A variable whose numerical value is determined by the outcome of a random experiment is called a* random variable.
>
> Definition 6·2: *Let* X *be a random variable which has the value set* $x_1, x_2, \ldots, x_k$ *with corresponding probabilities* $f(x_1), f(x_2), \ldots, f(x_k)$. *This value set together with the associated probabilities forms the* probability function of X.

In general then a random variable X is a rule by which every outcome E_i in the sample space is assigned a number. This number x_i is referred to as the value of the random variable. Consider an experiment with n outcomes and a r.v. X which takes on $k \le n$ values.

Sample Space of the Experiment Sample Space of the Random Variable

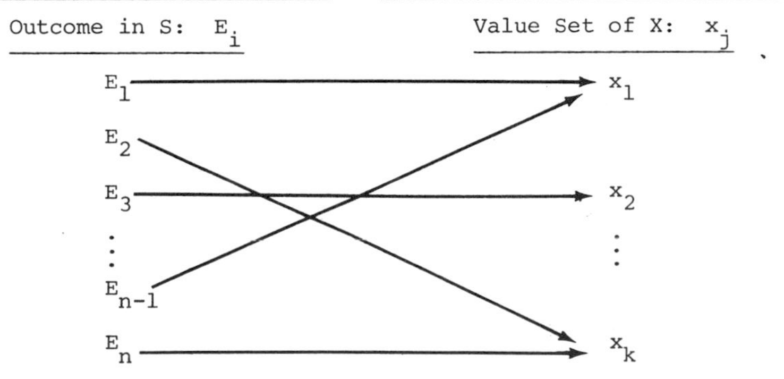

The above diagram shows how the sample space of the experiment is mapped on to the sample space of the r.v.--a many-to-one mapping.

Then the probability function of X is determined as

$$f(x_j) = \Pr(X = x_j) = \Sigma \Pr(E_i) \qquad j = 1, 2, \ldots, k$$

with the summation over all i such that $X = x_j$.

Probability functions have the following properties:

$$0 \le f(x_j) \le 1 \quad \text{for all } j$$

and

$$\sum_{j=1}^{k} f(x_j) = 1.$$

Example 1: Given the following p.f. for a r.v. X

x	0	1	2	3
f(x)	0	c	c	$3c^2$

determine c so that f(x) is a proper p.f. for x = 0, 1, 2 and 3.

Solution: Using the second property given above, we know that

$$0 + c + c + 3c^2 = 1,$$

which implies that

$$3c^2 + 2c - 1 = 0.$$

Factoring the l.h.s., we obtain

$$(3c - 1)(c + 1) = 0;$$

hence,

$$c = \frac{1}{3} \quad \text{or} \quad c = -1$$

are possible solutions. If, however, c = -1, then f(x) will be negative for f(1) and f(2) and greater than one for f(3). In these cases, the first property would not hold. Thus, if f(x) is to be a proper p.f. for x = 0, 1, 2, 3, then c must equal 1/3.

In the following discussion we may refer to probability functions as probability distributions. We will use the terms synonymously. The term probability distribution has arisen since we are interested in describing how the probability has been spread or distributed over the sample space.

We will now illustrate the techniques of determining probability functions.

Example 2: A balanced die is rolled. Let the r.v. X equal the face that appears. What is the p.f. of X?

Solution: This is a trivial example of mapping the sample space on to the space of the r.v. X. Here the correspondence is one-to-one. We easily see that the p.f. of X is

$$f(x) = \frac{1}{6} \qquad \text{for } x = 1, 2, 3, 4, 5, 6.$$

Example 3: Suppose a deck of cards consists of three cards, identical except for color (red, black and green). The cards are shuffled and dealt at random one card into each of three containers. If the containers are also red, black and green, what is the p.f. of X where

X = number of times the color of the card matches the color of the container.

Solution: A possible sample space would be the 3! = 6 permutations of the colors red, black and green:

$$S = \{(RGB), (RBG), (BGR), (BRG), (GBR), (GRB)\}.$$

Since the cards are dealt at random, each point is equally likely. Suppose the containers are arranged in the order red, black and green. Then for the sample point (RGB) there is one correct match, namely red. The sample space can be mapped on the space of the random variable X:

E_i	$Pr(E_i)$	x
RGB	$\frac{1}{6}$	1
RBG	$\frac{1}{6}$	3
BGR	$\frac{1}{6}$	0
BRG	$\frac{1}{6}$	1
GBR	$\frac{1}{6}$	1
GRB	$\frac{1}{6}$	0

We can then write the p.f. of X as

x	0	1	3
f(x)	$\frac{1}{3}$	$\frac{1}{2}$	$\frac{1}{6}$

Example 4: A very simple but frequently used type of random variable is one whose value set is {0, 1}. This type of random variable is called an *indicator random variable*. Let I be an indicator of the event E. Then I = 1 whenever E occurs and Pr(I = 1) = Pr(E); I = 0 when E does not occur and Pr(I = 0) = Pr($\bar{E}$). (a) If Pr(E) = p, find the p.f. of I. (b) If the r.v. J has the p.f.

j	0	1
Pr(J = j)	p	1 - p

is the r.v. J an indicator random variable? If so, for what event?
Solution: (a) Clearly, the p.f. of I can be summarized as

i	0	1
Pr(I = i)	1 - p	p

(b) Since the r.v. J takes on just two values 0 and 1, it is clearly an indicator. To determine what event it is "indicating," we look at the probability corresponding to J = 1. In this case the probability is 1 - p, which is Pr($\bar{E}$). Hence, J is an indicator for the event $\bar{E}$.

6-2 Cumulative Distribution Function

Frequently we may be interested in finding the probability of the event $\{X \le x\}$. This probability depends upon x and hence it is a function which changes its value as x changes. We shall call this probability the (cumulative) *distribution function* of X and represent it by F(x). Formally we have

$$F(x) = \sum_{t \le x} f(t),$$

where f(t) is the p.f. of X. To find the distribution function (d.f.), we add the probabilities for all values of the r.v. X which are less than or equal to a specified value x.

The distribution function can be represented graphically:

Suppose a r.v. X takes on six values $x_1, x_2, \ldots, x_6$. Then its distribution function can be graphed as:

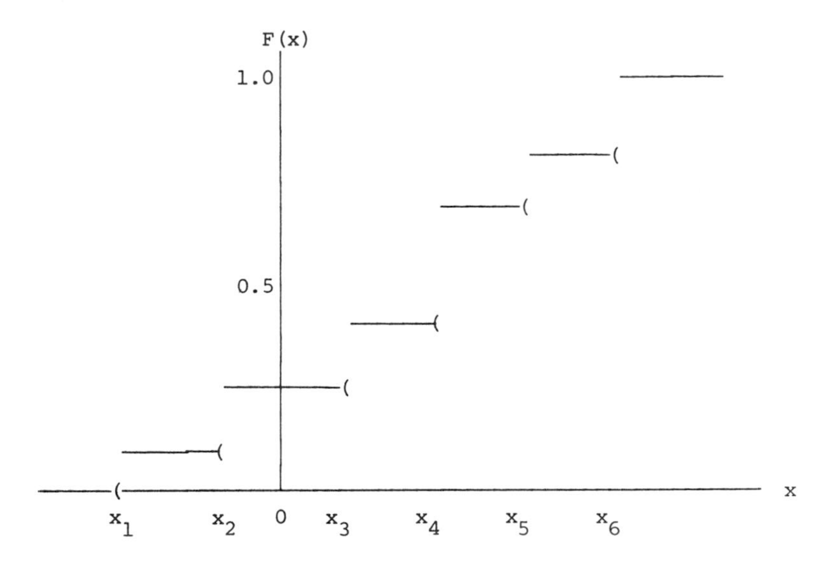

This is a "step-function" (resembling a staircase) with jumps or steps at $x_1, x_2, \ldots, x_6$. The jump at x_i is equal to the probability $f(x_i)$ at that point. The total height of the step at x_i is $f(x_1) + f(x_2) + \cdots + f(x_i)$. Between the jumps the function F(x) is a constant.

From the graph it is clear that F(x) is a nondecreasing function. What are its bounds? Suppose that m is the smallest value and M is the largest value in the value set of X. Then if $x < m$, the event $X \leq x$ is the null set; hence,

$$F(x) = \Pr(X \leq x) = 0 \quad \text{for } x < m.$$

In contrast if $x \geq M$, then the event $X \leq x$ is the entire sample space and

$$F(x) = \Pr(X \leq x) = 1 \quad \text{for } x \geq M.$$

Returning to our example of the number of males in families of three children, we can write the distribution function of X in a tabular form:

x	0	1	2	3
F(x)	$\frac{1}{8}$	$\frac{1}{2}$	$\frac{7}{8}$	1

or as a graph

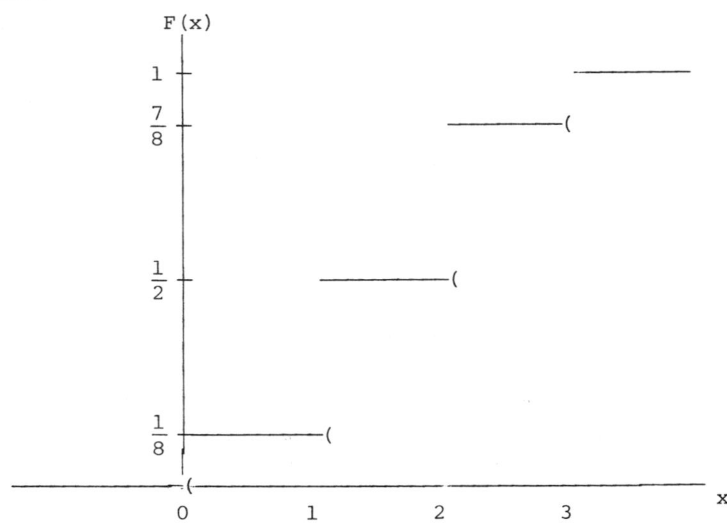

At times it may be more useful to express the distribution function as a mathematical formula:

$$F(x) = \sum_{t=0}^{x} \binom{3}{t} \left(\frac{1}{2}\right)^3 .$$

How can we use the distribution function to find the probability that X lies in an interval?

(i) $\Pr(a < X \leq b) = F(b) - F(a)$

(ii) $\Pr(a \leq X \leq b) = F(b) - F(a) + f(a)$

(iii) $\Pr(a \leq X < b) = F(b) - F(a) + f(a) - f(b)$

(iv) $\Pr(a < X < b) = F(b) - F(a) - f(b).$

Proof of these properties is left as an exercise; however, the following example illustrates their use.

Let the r.v. X have distribution function given by

$$F(x) = \begin{cases} 0 & x < -1 \\ \frac{1}{4} & -1 \le x < 1 \\ \frac{1}{2} & 1 \le x < 2 \\ \frac{2}{3} & 2 \le x < 3 \\ 1 & x \ge 3 \end{cases}$$

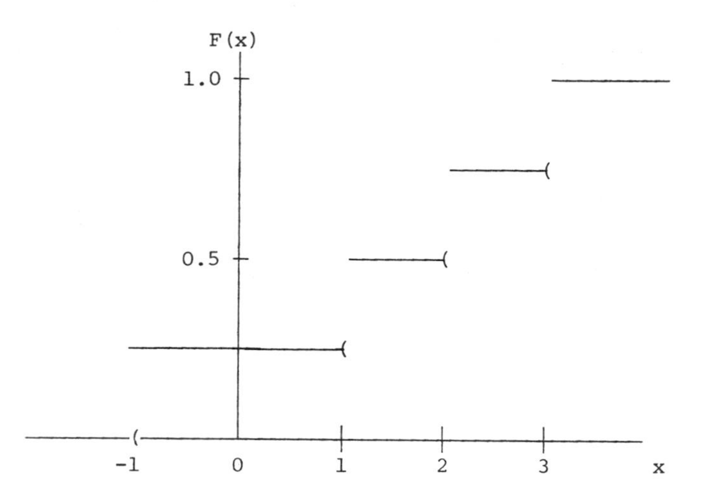

Using the above properties,

(i) $\Pr(-1 < X \le 2) = F(2) - F(-1) = \frac{2}{3} - \frac{1}{4} = \frac{5}{12}$

(ii) $\Pr(-1 \le X \le 2) = F(2) - F(-1) + f(-1) = \frac{2}{3} - \frac{1}{4} + \frac{1}{4} = \frac{2}{3}$

(iii) $\Pr(-1 \le X < 2) = F(2) - F(-1) + f(-1) - f(2)$

$$= \frac{2}{3} - \frac{1}{4} + \frac{1}{4} - \frac{1}{6} = \frac{1}{2}$$

(iv) $\Pr(1.5 < X < 2.7) = F(2.7) - F(1.5) - f(2.7)$

$$= \frac{2}{3} - \frac{1}{2} + 0 = \frac{1}{6}.$$

6-3 Functions of a Random Variable

In our example concerning the sex distribution of three-children families, we have defined the r.v.:

$$X = \text{number of male children.}$$

Now suppose your friend tells you that he is interested in the number of females. Can you find the p.f. of

$$Y = \text{number of females}$$

if you know the p.f. of X? We will make use of the mathematical relationship

$$Y = 3 - X.$$

We see that for the sample point (F, F, F), X = 0 which implies that Y = 3. Although the value of the r.v. has changed, the probability associated with the outcome is still 1/8. Thus we can tabulate the p.f. of Y and the p.f. of X as

h(y) or f(x)	$\frac{1}{8}$	$\frac{3}{8}$	$\frac{3}{8}$	$\frac{1}{8}$
x	0	1	2	3
y	3	2	1	0

Similarly, we might have been interested in the proportion of males

$$Z = \frac{X}{3}.$$

Then Z has the value set {0, 1/3, 2/3, 1} with corresponding probabilities 1/8, 3/8, 3/8, 1/8.

Example 5: Let I be an indicator r.v. for the event A with Pr(A) = p. Find the p.f. of I^2. Is I^2 an indicator? If so, for what event is it an indicator?

Solution: The p.f. of I is given by

i	0	1
$\Pr(I = i)$	$1 - p$	p

Let $J = I^2$. We see that the value set for J is the same as that of I and the p.f. of J is

j	0	1
$\Pr(J = j)$	$1 - p$	p

An examination of the above p.f. shows that $J = 1$ with probability $p = \Pr(A)$; thus J is an indicator of A. Here the r.v.'s I and I^2 are identical. In fact, this result can be generalized to any positive, integral power: I and I^k are identical r.v.'s and hence "indicate" the same event.

These examples have illustrated the relatively simple situation in which the transformation of variables is one-to-one. The following example describes a many-to-one transformation.

Example 6: Suppose that the outcomes of a medical diagnostic test can be coded as -1, 0, +1. The -1 and +1 responses represent different types of departure from normalcy. The relative frequencies of the responses are .24, .48, .28, respectively. The researcher then decides that he is interested only in "normal" and "abnormal" responses. Define an appropriate random variable and find its probability function.

Solution: Let X represent the outcomes of the diagnostic test. Then the p.f. of X is

x	-1	0	+1
$f(x)$	.24	.48	.28

The new r.v. Y measures only two responses:

$$Y = X^2.$$

The p.f. of Y is

y	0	1
g(y)	.48	.52

Note that Pr(Y = 1) = Pr(X = 1 or X = -1)--two points in X map on to one point in Y.

6-4 Joint Probability Functions

Although thus far we have restricted ourselves to a single r.v. defined on the sample space, we shall see that several r.v.'s may be of interest in the same experiment. We shall study the joint behavior of several random variables defined on the same sample space. Most of our discussion will be restricted to two random variables, but all the results can be easily generalized to more than two.

Example 7: A balanced die is rolled twice. (a) Describe an appropriate probability model for this experiment. (b) Define two r.v.'s:

F = Maximum of the two numbers

G = Number of times an even number appears

Find the joint p.f. of F and G.

Solution: (a) The sample space can be depicted as a two-dimensional array:

		Outcome on first roll					
		1	2	3	4	5	6
	1	1, 1	2, 1	3, 1	4, 1	5, 1	6, 1
	2	1, 2	2, 2	3, 2	4, 2	5, 2	6, 2
Outcome on	3	1, 3	2, 3	3, 3	4, 3	5, 3	6, 3
second roll	4	1, 4	2, 4	3, 4	4, 4	5, 4	6, 4
	5	1, 5	2, 5	3, 5	4, 5	5, 5	6, 5
	6	1, 6	2, 6	3, 6	4, 6	5, 6	6, 6

Assuming a uniform model (or independent rolls with constant probability), each point in S is equally likely with probability 1/36. (b) The r.v.'s F and G each represents a quantification of S with value sets {1, 2, 3, 4, 5, 6} and {0, 1, 2}, respectively. How can we determine the joint probability function of F and G from a knowledge of the probability model? Just as we did for a single r.v., we will determine the value of F and G as an ordered pair for each point in S. For example, if we consider the events (1, 3), (3, 3) and (3, 1) in S, they each correspond to F = 3 and G = 0. To find the probability that {F = 3 and G = 0}, we add the probabilities associated with these m.e. points in S:

$$\Pr(F = 3 \text{ and } G = 0) = \Pr\{(1, 3) \text{ or } (3, 3) \text{ or } (3, 1)\} = \frac{3}{36} .$$

Some outcomes in terms of F and G are not possible in S; thus

$$\Pr(F = 2 \text{ and } G = 0) = 0.$$

In summary, we have the joint probability function of F and G as

g \ f	1	2	3	4	5	6
0	$\frac{1}{36}$	0	$\frac{3}{36}$	0	$\frac{5}{36}$	0
1	0	$\frac{2}{36}$	$\frac{2}{36}$	$\frac{4}{36}$	$\frac{4}{36}$	$\frac{6}{36}$
2	0	$\frac{1}{36}$	0	$\frac{3}{36}$	0	$\frac{5}{36}$

Formally, then, the probability function of two random variables can be defined as

> Definition 6·3: *The* joint probability function *of two random variables* X *and* Y *is the sum of all probabilities in* S *at which* {X = x *and* Y = y}. *Symbolically we have*
>
> $$p(x, y) = \Pr(X = x \text{ and } Y = y)$$
>
> *for all ordered pairs of* x *and* y.

6-5 Marginal Probability Functions

We have seen how the joint and individual p.f.'s can be constructed directly from the sample space. Now suppose we know the joint p.f. p(x, y), which describes how the random variables X and Y vary jointly. Do we have sufficient information to obtain the individual p.f. of X and of Y?

In Example 7 we wish to find Pr{F = f} for f = 1, 2, ..., 6. Recall that the event {F = 1} can be written as

$$\{F = 1\} = \{(F = 1 \text{ and } G = 0) \text{ or } (F = 1 \text{ and } G = 1) \text{ or } (F = 1 \text{ and } G = 2)\},$$

where the r.h.s. events are m.e. Thus the

$$\Pr\{F = 1\} = \Pr(F = 1 \text{ and } G = 0) + \Pr(F = 1 \text{ and } G = 1) + \Pr(F = 1 \text{ and } G = 2) = \frac{1}{36} + 0 + 0 = \frac{1}{36}.$$

In general, then,

$$p_1(f) = \Pr(F = f) = \sum_{g=0}^{2} \Pr(F = f \text{ and } G = g)$$

for f = 1, 2, ..., 6; that is, the individual p.f. of F is found by adding over the rows of the array.

Similarly, if we add over the columns, we obtain the individual p.f. of G as

$$p_2(g) = \Pr(G = g) = \sum_{f=1}^{6} \Pr(F = f \text{ and } G = g)$$

for g = 0, 1, 2.

Since these individual p.f.'s can be depicted as an augmented row and column in the array ("in the margin," so to speak), we often refer to them as the marginal p.f. of F and G, respectively. The following table shows the marginal p.f.s

g \ f	1	2	3	4	5	6	$p_2(g) = \Pr(G = g)$
0	$\frac{1}{36}$	0	$\frac{3}{36}$	0	$\frac{5}{36}$	0	$\frac{9}{36}$
1	0	$\frac{2}{36}$	$\frac{2}{36}$	$\frac{4}{36}$	$\frac{4}{36}$	$\frac{6}{36}$	$\frac{18}{36}$
2	0	$\frac{1}{36}$	0	$\frac{3}{36}$	0	$\frac{5}{36}$	$\frac{9}{36}$
$p_1(f) = \Pr(F = f)$	$\frac{1}{36}$	$\frac{3}{36}$	$\frac{5}{36}$	$\frac{7}{36}$	$\frac{9}{36}$	$\frac{11}{36}$	

In general, then, we define the marginal probability function as

Definition 6·4: *Let* X *and* Y *be two random variables with joint p.f. given by* p(x, y). *Then the* marginal probability functions *of* X *and of* Y *are given by*

$$p_1(x) = \sum_y p(x, y) \qquad \textit{for all } x$$

and

$$p_2(y) = \sum_x p(x, y) \qquad \textit{for all } y.$$

6-6 Independence

In certain situations we do not obtain any added information about the random variables X and Y by knowing their joint p.f.. If a full knowledge of the joint p.f. is obtained simply by knowing the individual (marginal) probability functions, then the random variables X and Y are said to be independent.

Definition 6·5: *Two random variables* X *and* Y *with joint p.f.* p(x, y) *are said to be independent if and only if* $p(x, y) = p_1(x)p_2(y)$ *for each* (x, y).

This definition follows from the definition of independent events. The joint p.f. is simply a designation for the probability of intersecting events. Thus

$$p(x, y) = \Pr(X = x \text{ and } Y = y) = \Pr(X = x)\Pr(Y = y)$$

if and only if X and Y are independent.

In reference to Example 7, are the r.v.'s F and G independent? They are dependent since

$$p(1, 0) \neq p_1(1)p_2(0)$$

To further illustrate the concept of independence, let us again consider the sample space of Example 7 and define the r.v.'s"

X = outcome on first roll

Y = outcome on second roll.

The joint p.f. of X and Y is

$$\Pr(X = x \text{ and } Y = y) = \frac{1}{36}$$

for x = 1, 2, ..., 6; y = 1, 2, ..., 6. From basic considerations we know that the marginal p.f.'s are

$$\Pr(X = x) = \frac{1}{6} \qquad \text{for } x = 1, 2, \ldots, 6$$

and

$$\Pr(Y = y) = \frac{1}{6} \qquad \text{for } y = 1, 2, \ldots, 6.$$

Since

$$\Pr(X = x \text{ and } Y = y) = \Pr(X = x)\Pr(Y = y)$$

for all x and y, the r.v.'s X and Y are clearly independent.

Example 8: Suppose I_1 and I_2 are independent indicator r.v.'s each with p.f.:

$$\Pr(I_1 = 1) = \Pr(I_2 = 1) = p$$

$$\Pr(I_1 = 0) = \Pr(I_2 = 0) = 1 - p.$$

(a) Construct the p.f. of the sum $I_1 + I_2$. (b) Construct the p.f. of the product I_1I_2. (c) Show that the r.v. I_1I_2 is an indicator r.v. What event does it indicate?

Solution: (a) Let $J = I_1 + I_2$. Then the value set for J is {0, 1, 2}; hence J is <u>not</u> an indicator r.v. The event J = 0 occurs when both I_1 and I_2 are zero; hence,

$$\Pr\{J = 0\} = \Pr\{I_1 = 0 \text{ and } I_2 = 0\}.$$

Since I_1 and I_2 are independent,

$$\Pr(J = 0) = (1 - p)^2.$$

Similarly, J = 1 if either I_1 or I_2 (but not both) equals 1; that is,

$$\Pr\{J = 1\} = \Pr\{(I_1 = 0 \text{ and } I_2 = 1) \text{ or } (I_1 = 1 \text{ and } I_2 = 0)\}.$$

Again, using independence and the fact that the r.h.s. events are m.e., we have

$$\Pr(J = 1) = 2p(1 - p).$$

Using the same reasoning we have

$$\Pr(J = 2) = p^2.$$

In summary, the p.f. of J is

j	0	1	2
Pr(J = j)	$(1 - p)^2$	$2p(1 - p)$	p^2

(b) Let $H = I_1 I_2$ with value set {0, 1}; hence, H is an indicator. Using arguments similar to part (a), we get

h	0	1
Pr(H = h)	$1 - p^2$	p^2

(c) If I_1 indicates E_1 and I_2 indicates E_2, then H indicates $E_1 \cap E_2$ with probability p^2.

Example 9: Refer to the experimental set-up of Example 5 in Chapter 4. Let the r.v. X_i indicate a red marble on the ith draw. Are the r.v.'s X_1 and X_2 independent if the sampling is (a) WR? (b) WOR?

Solution: In Chapter 4 we have seen that regardless of whether the sampling is WR or WOR, the marginal p.f.'s of X_1 and X_2 will be identical:

$$\left.\begin{aligned} \Pr(X_i = 1) &= \frac{1}{3} \\ \Pr(X_i = 0) &= \frac{2}{3} \end{aligned}\right\} \quad \text{for } i = 1, 2$$

(a) The joint p.f. of X_1 and X_2 using WR

x_1 \ x_2	1	0	$p_1(x_1)$
1	$\left(\frac{1}{3}\right)^2$	$\left(\frac{1}{3}\right)\left(\frac{2}{3}\right)$	$\frac{1}{3}$
0	$\left(\frac{2}{3}\right)\left(\frac{1}{3}\right)$	$\left(\frac{2}{3}\right)^2$	$\frac{2}{3}$
$p_2(x_2)$	$\frac{1}{3}$	$\frac{2}{3}$	

(b) The joint p.f. of X_1 and X_2 using WOR

x_1 \ x_2	1	0	$p_1(x_1)$
1	$\left(\frac{3}{9}\right)\left(\frac{2}{8}\right)$	$\left(\frac{3}{9}\right)\left(\frac{6}{8}\right)$	$\frac{1}{3}$
0	$\left(\frac{6}{9}\right)\left(\frac{3}{8}\right)$	$\left(\frac{6}{9}\right)\left(\frac{5}{8}\right)$	$\frac{2}{3}$
$p_2(x_2)$	$\frac{1}{3}$	$\frac{2}{3}$	

In case (a), the joint p.f. is equal to the product of the marginal p.f.'s:

$$p(x_1, x_2) = p_1(x_1)p_2(x_2)$$

for $x_1 = 0, 1$ and $x_2 = 0, 1$. Thus X_1 and X_2 are <u>independent</u>. In contrast for case (b), $p(1, 1) \neq p_1(1)p_2(1)$; hence X_1 and X_2 are <u>not independent</u>.

In this example we have seen that, although the marginal p.f.'s for WR and WOR sampling are identical, the joint p.f.'s are <u>not</u>. In case (a) we can obtain the joint p.f. by knowing the marginal p.f.'s. When, however, X_1 and X_2 are not independent as in case (b), a knowledge of their marginals does not provide sufficient information for obtaining the joint p.f.

In summary, the following diagram may be useful in understanding the relationship between joint and marginal p.f.'s:

X and Y <u>independent</u>: joint p.f. $\rightleftarrows$ marginal p.f.

X and Y <u>dependent</u> : joint p.f. $\longrightarrow$ marginal p.f.

Problems

1. A coin is weighted so that Pr(H) = 3/8 and Pr(T) = 5/8. The coin is tossed until a head or three tails appear.
   (a) Describe a suitable sample space for this experiment and assign appropriate probabilities to each outcome. What assumptions have you made?
   (b) Let X be the number of tosses needed. Find the p.f. and d.f. of X.
   (c) Graph both the p.f. and d.f. obtained in part (b).

2. A hat check girl returns three hats at random to the three customers who had previously checked them. Smith, Jones and Brown, in that order, each receives one of the hats.
   (a) Discuss an appropriate probability model for this experiment.
   (b) Find the p.f. of M, the number of matches.
   (c) If there are n hats to be returned to n men, what is Pr(M = n)? Pr(M = n - 1)? Discuss these results.

3. Suppose W has a p.f. given by

w	-1	0	1
Pr(W = w)	3C	3C	6C

   (a) Determine C.
   (b) What is the p.f. of X = 2W + 1?

4. A tetrahedron is tossed into the air and the bottom face on which it comes to rest is noted. Each of the four faces--numbered 1, 2, 3 and 4--has an equal probability of being on the bottom when the tetrahedron comes to rest. Suppose the tetrahedron is tossed twice.
   (a) What is an appropriate sample space for this experiment? What is the probability associated with each outcome? Discuss carefully any assumptions which you have made in assigning these probabilities.

(b) Define a r.v. X to be the sum of the outcomes on the two tosses. Find the p.f. of X.

5. Repeat Problem 4 assuming that the tetrahedron is weighted so that the probability associated with each face is proportional to the number on it.

6. A salesman has four different stores as indicated in the map below:

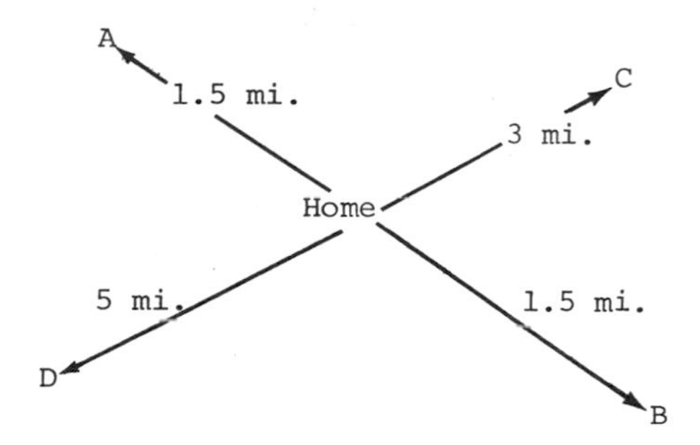

He decides to visit the store which telephones him first and then return home. The probability that the first call is from A is 1/6, from B is 1/3, from C is 1/3, from D is 1/6. Let X represent the distance to the first store and Y the total distance traveled to the store and back.

(a) Find the p.f. of X.

(b) Find the p.f. of Y.

7. Let X be a r.v. with d.f.

$$F(x) = \begin{cases} 0 & x < 4 \\ .1 & 4 \le x < 5 \\ .4 & 5 \le x < 6 \\ .7 & 6 \le x < 8 \\ .9 & 8 \le x < 9 \\ 1 & x \ge 9 \end{cases}$$

(a) Graph F(x).

(b) Find the p.f. of X and graph it.

(c) Using the d.f., determine

(i) $\Pr(X \le 6.5)$

(ii) $\Pr(X > 8.1)$

(iii) $\Pr(5 < X < 8)$

(iv) $\Pr(5 \le X < 8)$

8. Let X denote the number of hours you watch television on a randomly selected day. Suppose X has the following probability function where k is some constant:

$$\begin{aligned} f(x) &= .1 && x = 0 \\ &= kx && x = 1 \text{ or } 2 \\ &= k(6 - x) && x = 3 \text{ or } 4 \\ &= 0 && \text{otherwise} \end{aligned}$$

(a) Determine k so that f(x) is a proper p.f. of X.

(b) Graph f(x).

(c) Find $\Pr(X \ge 3)$.

9. Suppose three balanced coins are tossed.

(a) Let X be the number of tails. Find the p.f. of X.

(b) If Y is the number of heads minus the number of tails, find the p.f. of Y.

(c) Determine the joint p.f. of X and Y.

(d) Are X and Y independent?

10. Frequently in discussion randomness we may be interested in examining the number of *runs* which occur. An unbroken sequence of like outcomes is called a run. For example, in tossing a coin 10 times, we may observe

T H H H T T T T H H.

In this example there are four runs. Consider an analogous situation in which items coming off an assembly line are tested and found to be defective or nondefective. On a particular shift, five items are tested.

10. (a) If each item is equally likely to be defective or non-defective, discuss an appropriate sample space for the experiment.
    (b) Define X as the r.v. which counts the number of runs. Find the p.f. of X.
    (c) Graph the p.f. and d.f. of X.

11. A medical clinic is studying a new treatment. Suppose that each of five patients is given the new treatment and after one week the doctor diagnoses his condition as improved or not improved.
    (a) Discuss an appropriate probability model for this experiment if "improved" and "not improved" are equally likely.
    (b) If, after receiving the new treatment, the patients are twice as likely to improve, how does this affect the probability model developed in (a)?
    (c) Define a random variable X, the number of patients who improve. Find the p.f. of X using the assumptions in (1) part (a), (2) part (b).

12. Let X be a r.v. with probability function f(x) and cumulative distribution function F(x). Show that for any numbers x_1 and x_2 $(x_1 < x_2)$
    (a) $\Pr(x_1 < X \leq x_2) = F(x_2) - F(x_1)$
    (b) $\Pr(x_1 \leq X \leq x_2) = F(x_2) - F(x_1) + f(x_1)$
    (c) $\Pr(x_1 \leq X < x_2) = F(x_2) - F(x_1) + f(x_1) - f(x_2)$
    (d) $\Pr(x_1 < X < x_2) = F(x_2) - F(x_1) - f(x_2)$.

13. In an ESP experiment the subject attempts to match three different colored cards, red, black and white, with a control deck containing the same three cards.
    (a) If he arranges the cards at random and the correct ordering is red, black, white, find the p.f. of X, the number of matches.

(b) If the correct ordering is white, red, black, how will this affect the p.f. of X?

14. Suppose two balanced dice are rolled. Define two random variables:

M = maximum of the two numbers

S = sum of numbers appearing on the two dice

(a) Find the joint p.f. of (M, S).

(b) Construct the marginal p.f. of M and the marginal p.f. of S.

(c) Are M and S independent?

15. Suppose four distinguishable balls are randomly placed in three different boxes. Let X denote the number of balls in box 1 and Y denote the number of unoccupied boxes.

(a) Find the joint p.f. of (X, Y).

(b) Find the marginal p.f. of X and the marginal p.f. of Y.

(c) Determine whether or not X and Y are independent.

16. Two dead batteries have accidentally been placed in a box with six good batteries. Four batteries are drawn (WOR) from the box.

(a) Determine the p.f. of X, the number of dead batteries selected.

(b) Find the d.f. of X and graph it.

17. In certain societies male children are considered more desirable than females. In this situation husbands and wives may continue to have children until a son is born. In more recent times even such societies have been influenced by economic and sociological pressures to limit the family size to a maximum of three children.

(a) If the Pr(M) is p and Pr(F) = 1 - p, set up an appropriate probability model.

(b) Find the probability function of the number of girls in such families.

(c) Now consider families with three children. Find the p.f. of Y, the number of girls.

(d) Taking p = 1/2, compare the results you obtained in parts (b) and (c).

18. Suppose I_A and I_B are indicator r.v.'s for the events A and B, respectively; that is, $Pr(I_A = 1) = Pr(A)$ and $Pr(I_B = 1) = Pr(B)$.

(a) Find p.f.'s for each of the following random variables:

(i) $I_A I_B$

(ii) $I_A(1 - I_B)$

(iii) $(1 - I_A)(1 - I_B)$

(b) Using the results obtained in (a), show that each of these r.v.'s defined in (i), (ii) and (iii) is an indicator r.v. For what event is each an indicator?

(c) Show that the r.v.'s I_A and I_B are independent if and only if the events A and B are independent.

19. In the World Series, the American League (A) team and the National League (N) team play until one team wins four games. Suppose team A is stronger and has a probability 3/4 of winning each game independently of the outcomes on any other game.

(a) Discuss an appropriate sample space for this experiment. What probabilities would you assign for each point in S?

(b) Let X count the number of games in the series. Find the probability function of X.

20. In a sociological research project, families with 0, 1 and 2 children are being studied. Suppose that the relative frequency of each type is found to be

0 children, 30%

1 child, 40%

2 children, 30%.

A family is selected at random, and the family size (X) and the number of boys (Y) in the family are recorded. If Pr(boy) = Pr(girl) = 1/2,

(a) Construct the joint p.f. of (X, Y).

(b) Find the p.f. of X and the p.f. of Y.

(c) Are X and Y independent?

21. A number is selected at random from the set {3, 4, 5} and then a second number is selected from the set of all positive integers (not including zero) less than the first number selected.

    (a) Discuss an appropriate two-stage probability model for this experiment.

    (b) Define two random variables:

    X = first number selected

    Y = second number selected

    (i) Find the joint p.f. of (X, Y).

    (ii) Determine the marginal p.f.'s of X and of Y.

    (iii) Are X and Y independent?

22. Let X be a r.v. which takes on the values 0, 1, 2, 3 and 4 each with equal probability. Define a new r.v. $Y = X^3 \pmod 5$. Show that X and Y have the same p.f.'s.

23. Let X be a r.v. which takes on the values 0, 1, 2, 3, 4 each with equal probability.

    (a) Define a new r.v. $Y = X^2 \pmod 5$. What is the p.f. of Y?

    (b) Repeat this process; that is, let $Z = Y^2 \pmod 5$. Show that Z is an indicator r.v.

    (c) If we continue this process by forming an analogous r.v. from Z, what is its p.f.?

24. Let X and Y have the joint p.f. given by

$$p(x, y) = \frac{1}{66} xy \qquad x = 2, 4, 5; \quad y = 1, 2, 3.$$

Find the marginal p.f. of X and of Y.

25. A balanced coin is tossed four times. Consider two random variables:

$$X = \text{number of heads}$$

$$Y = \begin{cases} +1 & \text{if number of heads} > \text{number of tails} \\ 0 & \text{if number of heads} = \text{number of tails} \\ -1 & \text{if number of heads} < \text{number of tails} \end{cases}$$

(a) Construct the joint p.f. of X and Y.

(b) Determine whether or not X and Y are independent.

Chapter 7

Describing Random Variables and Their Distributions

7-1 Expectation

As a part of a program in mathematics a class of 50 children was given a quiz consisting of five questions. Let X be the number of questions answered correctly by each student. Clearly X can assume the values 0, 1, 2, 3, 4 or 5. How would you determine the average score in the class? When we consider the entire class, there are 50 values which X takes on. Some are obviously repetitions since X takes on only six distinct values. The quiz results might be summarized as:

Number correct	0	1	2	3	4	5
Number of students	2	2	6	20	15	5

We could obtain the average by adding the 50 individual scores and then dividing by 50. Due to the repetitions in the observations, the average can be obtained as a weighted average:

$$\text{Average} = (0)\frac{2}{50} + (1)\frac{2}{50} + (2)\frac{6}{50} + (3)\frac{20}{50} + (4)\frac{15}{50} + (5)\frac{5}{50} = 3.18.$$

Now suppose we had given this quiz to a large number of students. Then the observations might be summarized as:

Number correct	0	1	2	3	4	5
Number of students	n_0	n_1	n_2	n_3	n_4	n_5

where $\sum_{i=0}^{5} n_i = N$, the total number of students. Based on the concept of approximating probabilities by relative frequencies, we would say that the quantities n_0/N, n_1/N, ..., n_5/N would approach the probability of a child having 0, 1, ..., 5 correct answers, respectively, as N increases.

In an analogous manner, we can find the average value of a random variable X as the weighted mean of the possible values of X, each being weighted by the appropriate probability. This average value of X is called the (mathematical) expectation of X, the expected value of X or the mean of X. It may be symbolized as E(X), μ_X or μ. We can interpret this quantity in a physical sense as the center of gravity of the probability distribution of X--the balancing point of the distribution.

What do we mean by "mathematical expectation?" We are not using "expected" in the ordinary sense of the word. This value is "expected" in the sense that the long-run average over repeated experiments is likely to be close to it.

Definition 7·1: *Let* X *be a random variable with probability function*

x	x_1	x_2	...	x_k
f(x)	$f(x_1)$	$f(x_2)$	...	$f(x_k)$

Then the expected value of *X is given by*

$$E(X) = x_1 f(x_1) + x_2 f(x_2) + \cdots + x_k f(x_k) = \sum_{i=1}^{k} x_i f(x_i).$$

Example 1: Consider a r.v. V with p.f. given by

v	-5	-4	1	2
f(v)	$\frac{1}{4}$	$\frac{1}{8}$	$\frac{1}{2}$	$\frac{1}{8}$

What is the expected value of V?

Solution: $E(V) = (-5)\left(\frac{1}{4}\right) + (-4)\left(\frac{1}{8}\right) + (1)\left(\frac{1}{2}\right) + 2\left(\frac{1}{8}\right)$
$= -1.$

There is no implication that the expected value is frequent, highly probable or even possible. In this example the expectation of V does not correspond to any element in the value set of V.

Example 2: In reference to the matching distribution discussed in Example 3 of Chapter 6, find the mean value of X, the number of matches.

Solution: $\mu_X = E(X) = 0\left(\frac{2}{6}\right) + 1\left(\frac{3}{6}\right) + 3\left(\frac{1}{6}\right)$
$= 1.$

Example 3: Consider a r.v. I which is an indicator of the event E with Pr(E) = p. Show that E(I) = p.

Solution: Recall that the p.f. of I is

i	0	1
Pr(I = i)	1 - p	p

By definition $E(I) = 0(1 - p) + 1(p)$
$= p = Pr(E);$

that is, the expectation of an indicator is equal to the probability of the event which it is indicating.

Recall that in Section 6-3 we discussed functions of random variables and how to determine their p.f.'s. At this point you may wish to reread this section.

Now consider the question: What is the expected value of a function of a random variable?

Example 4: Let X be a r.v. with p.f.

x	-1	0	1
f(x)	.2	.3	.5

Find the E(2X).

Solution: Let Y = 2X. Then the p.f. of Y is

y	−2	0	2
Pr(Y = y)	.2	.3	.5

Using Definition 7·1, we have

$$E(Y) = E(2X) = -2(.2) + 0(.3) + 2(.5)$$
$$= .6 .$$

Example 5: In Example 6 of Chapter 6 we discussed a situation in which the r.v. X had a p.f.

x	−1	0	1
f(x)	.24	.48	.28

We then define a new r.v.

$$Y = X^2.$$

What is the expected value of X^2?

Solution: We know that the p.f. of $Y = X^2$ is

y	0	1
g(y)	.48	.52

hence

$$E(Y) = 0(.48) + 1(.52)$$
$$= .52.$$

If we first find the p.f. of the transformed r.v., it is quite easy to determine its expected value by using Definition 7·1. Can we, however, find the expected value of the transformed variable directly from the p.f. of the original r.v. X?

Definition 7·2: *Let* X *be a random variable with probability function*

x	x_1	x_2	···	x_k
f(x)	$f(x_1)$	$f(x_2)$	···	$f(x_k)$

and H *be a function of* X, *that is,* H(X). *Then the* expected value of the function H(X) *is*

$$E[H(X)] = H(x_1)f(x_1) + H(x_2)f(x_2) + \cdots + H(x_k)f(x_k)$$
$$= \sum_{i=1}^{k} H(x_i)f(x_i).$$

A quick redoing of Examples 4 and 5 will show that this formulation leads to the same solution. The proof of the equivalence of the two techniques is left as an exercise.

7-2 Laws of Expectation For A Single Random Variable

In this section we will develop properties of expectation. It will be useful to consider the expectation symbol E as a linear operator. We will see that the operator E behaves similarly to the summation operator Σ. Thus, before deriving the properties of expectation, you may find it helpful to review the rules of summation given in Appendix A.

Theorem 7·1: *Let* X *be a r.v. with p.f.* $f(x_i)$ *for* $i = 1, 2, \ldots, k$ *and a mean value* E(X). *Then*

$$E(aX + b) = aE(X) + b,$$

where a *and* b *are both constants.*

Proof:

By Definition 7·2

$$E(aX + b) = \sum_{i=1}^{k} (ax_i + b)f(x_i)$$

$$= \sum_{i=1}^{k} ax_i f(x_i) + \sum_{i=1}^{k} bf(x_i).$$

Since a and b are constants, the r.h.s. can be written as

$$a \sum_{i=1}^{k} x_i f(x_i) + b \sum_{i=1}^{k} f(x_i).$$

By Definition 7·1,

$$E(X) = \sum_{i=1}^{k} x_i f(x_i)$$

and the first property of p.f.'s,

$$\sum_{i=1}^{k} f(x_i) = 1.$$

Substituting these results, we obtain

$$E(aX + b) = aE(X) + b. \qquad \square$$

Theorem 7·1 is often referred to as expected value of a linear transformation of X. We will now consider several special cases of this theorem:

(i) If a = 0, then we have

$$E(b) = b;$$

that is, the expected value of a constant is that constant itself. Intuitively, this is reasonable since a constant has no random fluctuation.

(ii) If b = 0, then

$$E(aX) = aE(X).$$

In a physical sense we are saying that each value of the r.v. X has been multiplied by the same constant a; hence, we would expect the mean of the distribution of aX to be a times the mean of X.

All that we have done is to relabel the horizontal axis in our graph of the probability function.

(iii) If $a = 1$, then

$$E(X + b) = E(X) + b.$$

This case is quite similar to (ii) except that the origin has now been changed by addition of a constant b to each value of X. Reasonably one would expect the mean of the distribution to be translated b units.

(iv) $E(X - \mu) = 0$

This property follows directly from Theorem 7·1 by allowing $a = 1$ and $b = -\mu$. Using the expectation operator, we have

$$E(X - \mu) = E(X) - \mu.$$

Since E(X) is just another symbol for μ,

$$E(X - \mu) = 0.$$

This is a very important characterization of the mean: The expected value of the deviation of any r.v. from its mean is zero. The converse of this is also true; that is, if

$$E(X - d) = 0,$$

then d must be the mean value of X.

Example 6: Using the results of this section, find the E(Y) for Examples 4 and 5.

Solution: In Example 4 we find $E(X) = .3$; hence

$$E(Y) = E(2X) = 2E(X) = .6.$$

In contrast, although we know the $E(X) = 0$ in Example 5, we cannot determine $E(Y) = E(X^2)$. Why? Theorem 7·1 applies only to linear transformations of r.v.'s.

7-3 Variance

The mean or expected value of a random variable is useful in describing the center or balancing point of its distribution. Suppose, however, that the random variables X, Y, Z have the same expected value. Are you willing to say that their p.f.'s are identical?

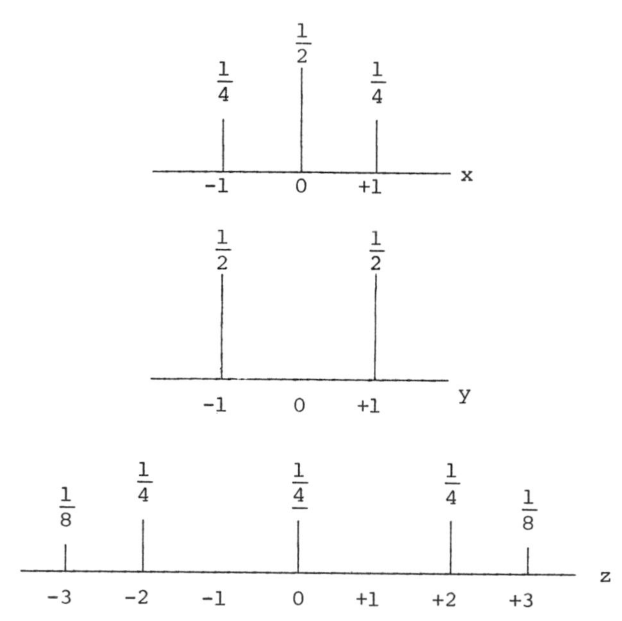

A glance at these three p.f.'s shows that although $E(X) = E(Y) = E(Z) = 0$, the distribution is different.

How might we measure the difference among these three? Many possible measures have been suggested. We will restrict ourselves to a quantity which measures the spread or variability about the mean. We would expect this measure to be small when the distribution is compact about the mean and large when the distribution is more spread out. We base this measure on the expected value of "squared deviations from the mean" and call it the *variance*.

Definition 7·3: *Let* X *be a r.v. with p.f.* $f(x_i)$ *for* $i = 1, 2, \ldots, k$ *and mean* $E(X) = \mu$. *Then the* variance of X, Var(X), *is defined by*

$$Var(X) = E(X - \mu)^2 = \sum_{i=1}^{k} (x_i - \mu)^2 f(x_i).$$

Note that this is just a special application of Definition 7·2, where $H(X) = (X - \mu)^2$. Sometimes the variance is also denoted by σ^2 or σ_X^2.

Since our measure is a function of squared deviations from the mean, it will measure the variability in squared units of X. Frequently it is useful to have a measure in the original units. To accomplish this, we take the positive square root of the variance.

Definition 7·4: *The* standard deviation of X *is the positive square root of the variance and is given by*

$$\sigma = \sigma_X = \sqrt{Var(X)}.$$

Example 7: Find the variance and standard deviation of each of the three random variables whose graphs are given at the beginning of this section.

Solution: In order to find the variance and standard deviation we must first determine the mean: $E(X) = E(Y) = E(Z) = 0$.

$$Var(X) = (-1 - 0)^2\left(\frac{1}{4}\right) + (0 - 0)^2\left(\frac{1}{2}\right) + (1 - 0)^2\left(\frac{1}{4}\right)$$
$$= \frac{1}{4} + \frac{1}{4} = \frac{1}{2}.$$

$$\sigma_X = \sqrt{.5} = .71$$

$$\mathrm{Var}(Y) = (-1 - 0)^2\left(\frac{1}{2}\right) + (1 - 0)^2\left(\frac{1}{2}\right)$$

$$= \frac{1}{2} + \frac{1}{2} = 1$$

$$\sigma_Y = 1.$$

$$\mathrm{Var}(Z) = (-3 - 0)^2\left(\frac{1}{8}\right) + (-2 - 0)^2\left(\frac{1}{4}\right) + (0 - 0)^2\left(\frac{1}{4}\right)$$

$$+ (2 - 0)^2\left(\frac{1}{4}\right) + (3 - 0)^2\left(\frac{1}{8}\right)$$

$$= \frac{9}{8} + 1 + 1 + \frac{9}{8} = 4.25$$

$$\sigma_Z = \sqrt{4.25} = 2.06.$$

As we would have expected from the graphs, the variances increase as the distributions become more spread out.

If the mean of X is not an integer, we may find that using Definition 7·3 to compute the variance is cumbersome. We shall now develop a simple computing formula:

> Computing formula for the variance:
>
> $$\mathrm{Var}(X) = E(X^2) - [E(X)]^2.$$

Proof:

By Definition 7·3

$$\mathrm{Var}(X) = E[X - E(X)]^2 = \sum_{i=1}^{k} [x_i - E(X)]^2 f(x_i).$$

Expanding the function on the r.h.s., we get

$$\mathrm{Var}(X) = \sum_{i=1}^{k} [x_i^2 - 2x_iE(X) + \{E(X)\}^2]f(x_i).$$

Removing the square brackets and using the rules of summation, we get

$$\text{Var}(X) = \sum_{i=1}^{k} x_i^2 f(x_i) - 2E(X) \sum_{i=1}^{k} x_i f(x_i) + \{E(X)\}^2 \sum_{i=1}^{k} f(x_i).$$

By definition,

$$E(X^2) = \sum_{i=1}^{k} x_i^2 f(x_i)$$

and

$$E(X) = \sum_{i=1}^{k} x_i f(x_i).$$

From the properties of p.f.'s, we know that

$$\sum_{i=1}^{k} f(x_i) = 1.$$

Making these substitutions, we have

$$\text{Var}(X) = E(X^2) - 2[E(X)][E(X)] + [E(X)]^2$$
$$= E(X^2) - [E(X)]^2.$$

□

Example 8: For the matching distribution, use the computing formula to find the variance of X, the number of matches.

Solution: We found $E(X) = 1$. Now

$$E(X^2) = 0^2\left(\frac{2}{6}\right) + 1^2\left(\frac{3}{6}\right) + 3^2\left(\frac{1}{6}\right) = 2.$$

Therefore,

$$\text{Var}(X) = E(X^2) - [E(X)]^2 = 2 - 1 = 1.$$

Example 9: Find the Var(I), where I is an indicator r.v.

Solution: If $Pr(E) = p$ and I is an indicator of E, then $E(I) = p$. Using the p.f. of I, we get

$$E(I^2) = 0^2(1 - p) + 1^2(p) = p.$$

Then

$$\text{Var}(I) = p - p^2 = p(1 - p).$$

The variance of an indicator r.v. is the product $\Pr(E)\Pr(\bar{E})$, the probability of the event it indicates times the probability of the complement event.

7-4 Laws of Variance for a Single Random Variable

In this section we will outline laws and properties of the variance (standard deviation) which are similar to those for the mean.

> Theorem 7·2: *Let* X *be a r.v. with p.f.* $f(x_i)$ *for* $i = 1, 2, \ldots, k$ *and mean* $E(X)$, *then*
>
> $$\mathrm{Var}(aX + b) = a^2 \mathrm{Var}(X),$$
>
> *where* a *and* b *are constants.*

Proof:

Recall that $E(aX + b) = aE(X) + b$. From Definition 7·3 we write

$$\mathrm{Var}(aX + b) = \sum_{i=1}^{k} \{ax_i + b - [aE(X) + b]\}^2 f(x_i).$$

Simplifying the r.h.s., we have

$$\sum_{i=1}^{k} [ax_i - aE(X)]^2 f(x_i) = a^2 \sum_{i=1}^{k} [x_i - E(X)]^2 f(x_i),$$

which, by definition of the variance of X, gives

$$\mathrm{Var}(aX + b) = a^2 \mathrm{Var}(X). \qquad \square$$

As a corollary to this theorem, we have the obvious result:

$$\sigma_{aX+b} = |a|\, \sigma_X.$$

This theorem for the variance of a linear function, like the one for expectation, results in several special cases:

(i) If a = 0,

$$\mathrm{Var}(b) = 0 \quad \text{and} \quad \sigma_b = 0.$$

Since b is a constant, it has no random variability.

(ii) If b = 0,

$$\mathrm{Var}(aX) = a^2\mathrm{Var}(X)$$

and

$$\sigma_{aX} = |a|\ \sigma_X.$$

The transformation aX represents either a shrinking or expansion of the scale of X depending upon whether a is less than or greater than one. Thus the variability will either decrease or increase.

(iii) If a = 1,

$$\mathrm{Var}(X + b) = \mathrm{Var}(X)$$

and

$$\sigma_{X+b} = \sigma_X.$$

The addition of a constant to every value of X does not affect the spread of the distribution--only the centering point is changed.

7-5 Chebyshev's Inequality

Having studied the mean and variance of random variables, we have seen that when the standard deviation is small, the probability piles up at the mean and when the standard deviation is large the probability tends to be spread out. In this section we will study a theorem due to the Russian mathematician, P. L. Chebyshev. This theorem provides us with further insight into the significance of the standard deviation as a measure of dispersion about the mean.

Before stating the theorem let us first consider a random variable X with probability function

x	0	1	2
f(x)	$\frac{5}{16}$	$\frac{8}{16}$	$\frac{3}{16}$

Let us determine the probability that X is *at or within*

(i) One standard deviation of the mean

(ii) Two standard deviations of the mean

(iii) Three standard deviations of the mean.

Noting that E(X) = 7/8 and σ = .6960, we have

(i) $\Pr\{ | X - \frac{7}{8} | \le .6960\} = \frac{8}{16} = .5$

(ii) $\Pr\{ | X - \frac{7}{8} | \le 2(.6960)\} = 1$

(iii) $\Pr\{ | X - \frac{7}{8} | \le 3(.6960)\} = 1.$

Theorem 7·3: *Let X be a random variable with finite mean μ and finite standard deviation σ. Then for any positive constant c, the probability that X lies outside of the closed interval $[\mu - c\sigma, \mu + c\sigma]$ is less than $1/c^2$; that is,*

$$\Pr\{ | X - \mu | > c\sigma\} < \frac{1}{c^2} .$$

Proof:

The value set of X can be divided into two regions:

I: Within and including the boundary of the closed interval $[\mu - c\sigma, \mu + c\sigma]$.

II: Beyond the closed interval

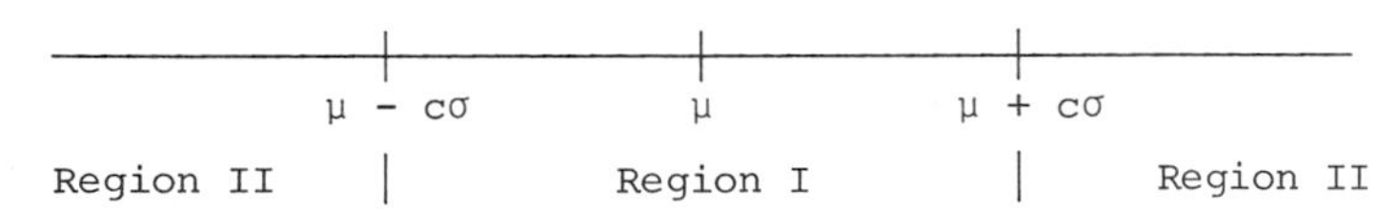

In region I, $| X - \mu | \le c\sigma$; while in region II, $| X - \mu | > c\sigma$. By definition

$$\text{Var}(X) = \sigma^2 = \sum_{i=1}^{k} (x_i - \mu)^2 f(x_i).$$

Since each term of this sum is nonnegative, omitting some terms cannot increase the value of the sum. If we delete the terms in region I, then

$$\sigma^2 \geq \Sigma^* (x_i - \mu)^2 f(x_i),$$

where Σ^* indicates summation over region II. If X takes on no values in region I, then the equality will be attained. Since in region II $|x_i - \mu| > c\sigma$, this sum can be further decreased by replacing each $(x_i - \mu)^2$ by the smaller value $c^2\sigma^2$; that is,

$$\sigma^2 > \Sigma^* c^2\sigma^2 f(x_i) = c^2\sigma^2\Sigma^* f(x_i).$$

But

$$\Sigma^* f(x_i) = \Pr\{|X - \mu| > c\sigma\}.$$

Hence

$$\sigma^2 > c^2\sigma^2 \Pr\{|X - \mu| > c\sigma\}.$$

Since $\sigma^2 > 0$, we have

$$\frac{1}{c^2} > \Pr\{|X - \mu| > c\sigma\}.$$

The probability assigned to values of X outside the interval $[\mu - c\sigma, \mu + c\sigma]$ is less than $1/c^2$. Or equivalently, the probability assigned to values of X within the interval is more than $1 - 1/c^2$; that is,

$$\Pr\{|X - \mu| \leq c\sigma\} > 1 - \frac{1}{c^2}. \qquad \square$$

If $0 < c \leq 1$, then the inequality does not really give any useful information. If $c > 1$, then Chebyshev's inequality does give some information about the probability contained in intervals about the mean. For example, for $c = 2$

$$\Pr\{|X - \mu| \leq 2\sigma\} > 1 - \frac{1}{4} = .750$$

and for c = 3

$$\Pr\{|X - \mu| \le 3\sigma\} > 1 - \frac{1}{9} = .889.$$

Now compare the exact results from the previous example. Each of the probabilities that we computed were greater than the smallest bound given above.

The probability statement given by Theorem 7·3 applies to <u>any</u> random variable. One pays a price for such great generality. As we saw in the example, one cannot expect an inequality which applies to all random variables to be definitive when dealing with a particular random variable. Nevertheless, Chebyshev's inequality is a powerful tool for dealing with proofs in the theory of probability.

Problems

1. Let X be a r.v. with p.f.

x	-2	-1	0	1	2
Pr(X = x)	.2	.1	.3	.3	.1

   (a) Find the E(X) and Var(X).
   (b) Find the p.f. of the r.v. Y = 3X - 1. Using the p.f. of Y, determine E(Y) and Var(Y).
   (c) Compare the answer you obtained in (b) with 3E(X) - 1 and 9Var(X).

2. Consider the two random variables X and Y with p.f.'s:

x	-1	0	1	2	3
Pr(X = x)	.125	.5	.20	.05	.125

y	-1	1	3	5	7
Pr(Y = y)	.125	.5	.20	.05	.125

   (a) Find E(X), E(Y), Var(X), Var(Y).
   (b) What is the relationship between X and Y?

(c) How are E(X) and E(Y) related?

(d) How are Var(X) and Var(Y) related?

3. In a certain industrial process the temperature never varies by more than 2°F from 77°F. The temperature X is a random phenomenon with p.f.

x	75	76	77	78	79
f(x)	$\frac{2}{10}$	$\frac{2}{10}$	$\frac{3}{10}$	$\frac{2}{10}$	$\frac{1}{10}$

(a) Let Y be a r.v. with expectation μ_Y and variance σ_Y^2. Recalling Theorems 7.1 and 7.2, find E(Z) and Var(Z) when Y = Z - c, where c is an arbitrary constant.

(b) Define a r.v. V = X - 77. Find E(V) and Var(V). Using the results of (a), find E(X) and Var(X). Note that this technique of "coding" the value set of the random variable makes the calculations much easier.

(c) Generalize parts (a) and (b) in the following way: Suppose a new r.v. is constructed using the rule Y = (Z - c)/d; that is, a constant c is subtracted from each number in the value set of Z and the result is then divided by some constant d. Find E(Z) and Var(Z) as functions of E(Y) and Var(Y).

(d) Suppose the temperature readings had been recorded on a Celsius scale: C = 5(F - 32)/9. Using the results of part (c), show how to find the mean and variance on the Fahrenheit scale.

4. A bag contains four red and six white marbles. Three marbles are chosen at random. Let X be the number of red marbles in the sample.

(a) Find the p.f. of X if the sampling is

  (i) WR.

  (ii) WOR.

(b) Suppose this experiment was actually performed 100 times. What would the expected frequencies of 0, 1, 2 and 3 be if the sampling was done

(i) WR?

(ii) WOR?

(c) If the actual experimental results were

Number of red	0	1	2	3
Frequency	15	52	29	4

<u>explain</u> which method of sampling you think was used.

5. Let X be a r.v. taking on the values x_i with probabilities $f(x_i)$ for $i = 1, 2, \ldots, k$. Show that if

$$\sum_{i=1}^{k} (x_i - c)f(x_i) = 0,$$

then $E(X) = c$. This is the converse of special case (iv). Theorem 7·1.

6. Show that $E[(X - c)^2] = \text{Var}(X) + [E(X) - c]^2$. For what value of c is $E[(X - c)^2]$ a minimum?

7. Let X be a r.v. with expected value and variance given by E(X) and Var(X). Now consider the r.v.

$$Y = a + bX + cX^2$$

where a, b and c are arbitrary constants. Show that

$$E(Y) = a + bE(X) + c[E(X)]^2 + c\text{Var}(X).$$

8. Suppose the p.f. f(x) is symmetrical about the line x = a; i.e., $f(a + x) = f(a - x)$ for all x. Show that $E(X) = a$.

9. Let X be a r.v. with p.f. $f(x_i)$ for $i = 1, 2, \ldots, k$. Prove, using Definition 7·3, that

(a) $\text{Var}(cX) = c^2\text{Var}(X)$

(b) $\text{Var}(X + d) = \text{Var}(X)$

where c and d are constants.

10. Suppose X is a r.v. with p.f. $f(x_i)$ for i = 1, 2, ..., k. Consider the following functions of X:

$$Y = |X - E(X)|$$
$$Z = \{X - E(X)\}^2.$$

(a) Prove that $E(Y) \geq 0$ and $E(Z) \geq 0$.

(b) Show that the equalities are attained if and only if $Pr[X = E(X)] = 1$. Describe the p.f. of X in this case.

(c) Why does it seem reasonable to say that the larger E(Y) and E(Z) are the more dispersed X is about E(X)?

(d) What special name do we attach to E(Z)?

11. Each of two wards has 100 voters. The number of Liberals in ward A is 20 and in ward B is 60. Several methods have been suggested for selecting two voters at random:

(i) Select one person at random from each ward.

(ii) Select two people at random (WR) from the entire set of 200 voters.

(iii) Select two people at random (WOR) from the entire set of 200 voters.

(iv) Select one of the two wards at random and then select two voters (WR) from the selected ward.

(v) Select one of the two wards at random and then select two voters (WOR) from the selected ward.

(a) Find the distribution of S, the number of Liberals for each of the five methods.

(b) What are the mean and variance for each distribution?

12. An unbalanced coin with Pr(H) = p and Pr(T) = 1 - p is tossed until a head appears or three times. If X is the number of tosses needed, find the

(a) Distribution of X.

(b) Mean of X.

(c) Variance of X.

13. Suppose that we do not know the p.f. of X. In this case the expected value of X can be found directly from the probability model $[E_i, Pr(E_i)]$ for i = 1, 2, ..., k:

$$E(X) = X(E_1)Pr(E_1) + \cdots + X(E_k)Pr(E_k),$$

where $X(E_i)$ is value of X associated with outcome E_i. Show that this is equivalent to Definition 7·1:

$$E(X) = \sum_{j=1}^{k} x_j f(x_j).$$

14. Let X be a r.v. which takes on the values $x_1 \le x_2 \le \cdots \le x_k$ with corresponding probabilities $f(x_1), f(x_2), \ldots, f(x_k)$. Show that

$$x_1 \le E(X) \le x_k.$$

When will the bounds of the inequalities be attained?

15. Let the mean and variance of the r.v. Z be 100 and 25, respectively; evaluate
    (a) $E(Z^2)$
    (b) Var(2Z + 100)
    (c) Standard deviation of 2Z + 100
    (d) E(−Z)
    (e) Var(−Z)
    (f) Standard deviation of (−Z)

16. Suppose that the probability that a voter will support new tax legislation in a referendum is p. A voter is selected at random. If he opposes, we let the r.v. X = 0; if he favors, X = 1.
    (a) Find E(X) and Var(X).
    (b) Show that Var(X) ≤ 1/4. For what value of p is Var(X) = 1/4?
    (c) Show that if X is any r.v. such that $E(X^2) = E(X)$, then Var(X) ≤ 1/4.
    (d) What kind of r.v. do we call X?

17. A box contains five slips of paper numbered 1, 2, 3, 4 and 5.
    (a) Let X denote the number on the slip of paper if one slip is drawn at random. Find the E(X) and Var(X).
    (b) Now suppose that two slips are drawn at random with replacement. Let Y be the maximum number appearing on the two slips. Find E(Y) and Var(Y).
    (c) Repeat part (b) using sampling without replacement.

18. In Problem 19 of Chapter 6 find the expected number of games in a series.

19. Consider a r.v. Z which takes on the integral values a to a + b with equal probabilities.
    (a) Find the p.f. of Z.
    (b) Show that the middle value a + b/2 is the expected value of Z.
    (c) Find the Var(Z).
    [Hint: Recall that:

$$\sum_{i=1}^{k} i = \frac{k(k+1)}{2} \quad \text{and} \quad \sum_{i=1}^{k} i^2 = \frac{k(k+1)(2k+1)}{6} \;]$$

20. Suppose a box contains n cards numbered 1, 2, ..., n. A card is drawn at random and the number on it is denoted by X.
    (a) Find the p.f. and d.f. of X. Draw a graph of each.
    (b) Determine E(X) and Var(X).
    (c) Give two examples in which the results of (a) and (b) are applicable.
    (d) How are the results in (b) related to those obtained in Problem 19?

21. A golf ball manufacturer is considering a new production process. In the new process the probability of a defective, which cannot be sold, is .05 while in the established process it is .08. Using the established process, the cost of production is 40 cents per ball while in the new process it is 60 cents.

21. The balls are sold at $1.25 each. If the manufacturer wishes to maximize his expected profit, which process should be used?

22. A pole light has three identical glass globes. If the three lights are removed for cleaning and then replaced at random, what is the expected number of globes that will be replaced in their original positions? (Hint: Use the results of Problem 13.)

23. Suppose one dead battery has been put into a box with three good ones. If the batteries are tested one at a time until the dead one is found, find the expected number of batteries which have been tested when the defective is found.

24. A lottery has a first prize of $100, a second prize of $50 and four $25 prizes. Would you pay $1 for a ticket if there were (a) 100 tickets, (b) 500 tickets being sold?

25. A game played by two people is said to be fair if the expected return for each player is zero.
    (a) Suppose Tom and Jerry roll a balanced die and Tom agrees to pay Jerry $5 if the score is less than 3. How much should Jerry pay Tom when the score is greater than or equal to 3 if the game is to be fair?
    (b) If the game in (a) is altered so that no one wins any money if a 3 appears, what should Jerry pay Tom so that the game will be fair?

26. Jack pays Bill $1 and two fair dice are rolled. Jack receives $2 from Bill if one 6 appears, $4 if two 6's appear and no return otherwise.
    (a) Find Jack's expected net gain. Is this a fair game?
    (b) What should Jack pay Bill as an entrance fee in order that the game be fair?

27. A box has five balls numbered from 1 to 5. You are offered two gambling options.

    (1) Pay \$1, draw a ball and then be paid the number of dollars appearing on the ball.

    (2) Pay \$1 and draw a ball. If the number on the ball is greater than 2, be paid the amount on the ball. If the number is less than or equal to 2, return the ball. Then pay \$2, draw another ball and be paid the amount on it.

    (a) Let Y_1 and Y_2 be your net gain when you use options (1) and (2), respectively. Determine the p.f. of Y_1 and of Y_2.

    (b) Are options (1) and (2) fair?

    (c) If you wish to maximize your expected net gain, which option should you accept?

28. Suppose two slot machines are available. Machine A costs \$1 to play and, with probability 1/3, returns \$2 and, with probability 2/3, returns nothing. Machine B also costs \$1 to play but, with probability 1/6, returns \$4 and, with probability 5/6, returns nothing. Consider the following two methods of play:

    (1) Play machine A first; if you win, play A again; otherwise switch to B.

    (2) Play machine B first; if you win, play B again; otherwise switch to A.

    (a) Find the mean and variance of your net gain under the two possible methods of play.

    (b) Which of the two methods would you prefer and why?

29. A box is filled with balls marked with the integers 1, 2, ..., n. For each integer k, there are k balls marked with k. A ball is drawn at random and its number is noted. Find the expected value and variance of the number chosen. (Hint: The following

may be useful:

$$\sum_{i=1}^{t} i = \frac{t(t+1)}{2}; \qquad \sum_{i=1}^{t} i^2 = \frac{t(t+1)(2t+1)}{6}$$

$$\sum_{i=1}^{t} i^3 = \frac{t^2(t+1)^2}{4})$$

30. The expected value of a random variable X is not necessarily the only measure of central tendency which we can use to describe the distribution of X. Other frequently used measures are the median and mode. The median of the distribution of X is any number, med, such that

$$\Pr\{X < \text{med}\} \le \frac{1}{2} \quad \text{and} \quad \Pr\{X > \text{med}\} \le \frac{1}{2}.$$

Show that this definition is equivalent to each of the following statements:

(a) $\Pr\{X < \text{med}\} \le \frac{1}{2} \le \Pr\{X \le \text{med}\}$.

(b) Let t be some number less than med and F(x) the distribution function, then

$$F(t) \le \frac{1}{2} \le F(\text{med}).$$

31. Note that in Problem 30 we used the term a median rather than the median because more than one number may satisfy this property. Construct an example in which a unique median does not exist.

32. The "most probable value" of X is called the mode. More formally, the mode is that value of the random variable at which the probability is the greatest. Like the median, the mode is not necessarily unique. Construct an example to show that the mode need not be a unique value of X.

33. Let T be the sum of the numbers when two fair dice are rolled. Find the mean, median and mode of T.

34. For each of the following random variables find the

$$\Pr\{| X - \mu_X | \le c\sigma\}$$

for c = 1.5, 2, 2.5, 3 and compare these with the corresponding probabilities given by Chebyshev's inequality:

(a) The number of points when a single balanced die is rolled.

(b) The sum when two fair dice are rolled.

(C) The number of tails in tossing three balanced coins.

35. In Theorem 7·3 we included the end points $\mu - c\sigma$ and $\mu + c\sigma$ as part of region I. Show that if these end points are not included in region I but are incorporated into the two parts of region II, then the corresponding probability satisfies

$$\Pr\{| X - \mu | \ge c\sigma\} \le \frac{1}{c^2}$$

36. Comment on the following statement: When $0 < c \le 1$, Chebyshev's theorem is not useful.

37. Let X be a random variable taking on the values -k, 0 and k with probabilities p, 1 - 2p and p, respectively.

(a) Determine E(X) and Var(X).

(b) Show that $k = \sigma$ when $p = 1/2$. Describe the probability function in this situation.

(c) If $k = 2\sigma$, what does p equal?

(d) If $k = 3\sigma$, what does p equal?

(d) By a proper choice of p, can you make $k = c\sigma$, for any positive c? If so, what is the proper choice of p in terms of c? What is the probability that $| X - \mu |$ is at least as great as $c\sigma$?

Chapter 8

Describing the Joint Behavior of Several Random Variables

In Chapter 6 we discussed how two or more random variables might be defined on the same sample space. We summarized their joint behavior by a joint probability function. At times the probability function may be a cumbersome description of the random variation. As a remedy we developed more concise measures--the mean and the variance. In this chapter we will introduce some concise measures involving more than one variable: covariance and correlation coefficient. We will also develop laws of expectation and variance involving several variables.

8-1 Expectation of a Function of Two Random Variables

Extending the concepts of the last chapter to several variables, we will now define the expected value of a function of two random variables.

Definition 8·1: *Let* X *and* Y *be two random variables with joint probability function* $p(x_i, y_j)$ *for* $i = 1, 2, \ldots, n$ *and* $j = 1, 2, \ldots, m$. *Consider a function* $G(X, Y)$. *Then the* *expected value of* $G(X, Y)$ *is*

$$E[G(X, Y)] = \sum_{i=1}^{n} \sum_{j=1}^{m} G(x_i, y_j)p(x_i, y_j).$$

Note that this is just a generalization of Definition 7·2. To illustrate this definition, consider the following example.

Example 1: Let X and Y have the joint p.f.

y \ x	0	1	2	3
1	$\frac{1}{8}$	0	0	$\frac{1}{8}$
2	0	$\frac{1}{8}$	$\frac{1}{8}$	0
3	0	$\frac{2}{8}$	$\frac{2}{8}$	0

What is the expected value of (a) XY? (b) X/Y?

Solution: (a) Here G(X, Y) = XY. Then

$$E(XY) = (0)(1)\frac{1}{8} + (3)(1)\frac{1}{8} + (1)(2)\frac{1}{8} + (2)(2)\frac{1}{8} + (1)(3)\frac{2}{8} + (2)(3)\frac{2}{8} = \frac{27}{8}.$$

(b) Similarly,

$$E\left(\frac{X}{Y}\right) = \left(\frac{0}{1}\right)\frac{1}{8} + \left(\frac{3}{1}\right)\frac{1}{8} + \left(\frac{1}{2}\right)\frac{1}{8} + \left(\frac{2}{2}\right)\frac{1}{8} + \left(\frac{1}{3}\right)\frac{2}{8} + \left(\frac{2}{3}\right)\frac{2}{8} = \frac{13}{16}.$$

Definition 8·1 can easily be extended for more than two random variables.

8-2 Covariance

In studying the behavior of two random variables simultaneously, we are faced with trying to find a measure of how the r.v.'s vary jointly. Essentially we are interested in generalizing the concept of "variance."

Definition 8·2: *The* covariance of two random variables *is the expected value of the product of the deviations* $X - E(X)$ *and* $Y - E(Y)$:

$$\text{Cov}(X, Y) = \sigma_{XY} = E\{[X - E(X)][Y - E(Y)]\}.$$

From Definition 8·2 it is obvious that the commutative law holds for covariance:

$$\text{Cov}(X, Y) = \text{Cov}(Y, X).$$

It is also clear that the covariance of a random variable with itself is its variance because

$$\text{Cov}(X, X) = E\{[X - E(X)][X - E(X)]\} = \text{Var}(X).$$

To actually calculate the covariance, we apply Definition 8·1 with $G(X, Y) = [X - E(X)][Y - E(Y)]$:

$$\text{Cov}(X, Y) = \sum_{x}\sum_{y} [x - E(X)][y - E(Y)]p(x, y),$$

where the summation extends over all possible pairs of values of x and y.

From this definition of Cov(X, Y) we see that the covariance is a measure of the extent to which the values of X and Y tend to increase or decrease together. If X has values greater than E(X) when Y has values greater than E(Y) and X has values less than E(X) when Y has values less than E(Y), then the Cov(X, Y) is positive. On the other hand, if X has values less than E(X) when Y has values greater than E(Y) or vice versa, then the Cov(X, Y) is negative. We shall see that if X and Y are independent, then the Cov(X, Y) must be zero. In contrast, a zero covariance does not imply that X and Y are independent. In summary, the Cov(X, Y) can be positive or negative and has no bounds:

$$-\infty \leq \text{Cov}(X, Y) \leq \infty.$$

As in the case of the variance, we will find that a computational formula for the covariance will be useful.

Computational Formula for Covariance:

$$\text{Cov}(X, Y) = E(XY) - E(X)E(Y).$$

Proof:

By definition,

$$\text{Cov}(X, Y) = E\{[X - E(X)][Y - E(Y)]\}$$

$$= \sum_{x}\sum_{y} [x - E(X)][y - E(Y)]p(x, y),$$

where p(x, y) is the joint p.f. and the double summation is over all possible pairs of x and y. Removing the square brackets on the r.h.s., we have

$$\text{Cov}(X, Y) = \sum_{x}\sum_{y} xyp(x, y) - E(X) \sum_{y} y\{\sum_{x} p(x, y)\}$$

$$- E(Y) \sum_{x} x\{\sum_{y} p(x, y)\} + E(X)E(Y) \sum_{x}\sum_{y} p(x, y).$$

From Chapter 6 we know that if we symbolize the marginal p.f. of X and Y, respectively, as $p_1(x)$ and $p_2(y)$, then

$$p_1(x) = \sum_{y} p(x, y); \qquad E(X) = \sum_{x} x\{\sum_{y} p(x, y)\}$$

$$p_2(y) = \sum_{x} p(x, y); \qquad E(Y) = \sum_{y} y\{\sum_{x} p(x, y)\}.$$

Substituting the above, we have

$$\text{Cov}(X, Y) = E(XY) - E(X)E(Y) - E(Y)E(X) + E(X)E(Y)$$

$$= E(XY) - E(X)E(Y). \qquad \square$$

We will illustrate this computational formula using the p.f. given in Example 1. First we will determine the marginal p.f.'s and the individual expectations of X and Y.

x	0	1	2	3
$p_1(x)$	$\frac{1}{8}$	$\frac{3}{8}$	$\frac{3}{8}$	$\frac{1}{8}$

$E(X) = 1.5$

y	1	2	3
$p_2(y)$	$\frac{2}{8}$	$\frac{2}{8}$	$\frac{4}{8}$

$E(Y) = 2.25$

Recall that in part (a) of Example 1 we determined the E(XY) to be 27/8; hence,

$$\text{Cov}(X, Y) = \frac{27}{8} - \left(\frac{3}{2}\right)\left(\frac{9}{4}\right) = 0.$$

Example 2: Given the joint p.f. of X and Y and the respective marginal p.f.'s, $p_1(x)$ and $p_2(y)$:

y \ x	1	2	3	4	5	$p_2(y)$
1	.1	0	0	0	0	.1
2	0	.2	0	.1	0	.3
3	0	0	.2	0	0	.2
4	0	.1	0	.2	0	.3
5	0	0	0	0	.1	.1
$p_1(x)$	.1	.3	.2	.3	.1	

Determine the Cov(X, Y).

Solution: A quick calculation shows that E(X) = E(Y) = 3. Then, using the joint p.f., we have

$$\begin{aligned} E(XY) &= (1)(1)(.1) + (2)(2)(.2) + (3)(3)(.2) \\ &\quad + (4)(4)(.2) + (5)(5)(.1) + (4)(2)(.1) \\ &\quad + (2)(4)(.1) = 10. \end{aligned}$$

Note that only seven of the possible 25 pairs of (x, y) contribute to the E(XY) since p(x, y) is zero for all others. Using the computation formula, we have

$$\text{Cov}(X, Y) = 10 - (3)(3) = 1.0.$$

Example 3: A balanced die is rolled twice. Let

$$X = \text{outcome of first roll}$$
$$Y = \text{outcome of second roll.}$$

Find the covariance between X and Y.

Solution: From our discussion in Section 6-6, we know that $E(X) = E(Y) = 3.5$. Recalling that

$$p(x, y) = \frac{1}{36} \quad \text{for } x = 1, 2, \ldots, 6 \text{ and } y = 1, 2, \ldots, 6,$$

we can write

$$E(XY) = \sum_{y=1}^{6} \sum_{x=1}^{6} \frac{xy}{36} = \frac{(21)(21)}{36} = 12.25.$$

Therefore,

$$\text{Cov}(X, Y) = 12.25 - (3.5)(3.5) = 0.$$

This example illustrates an important property of independent random variables.

> Theorem 8·1: *If* X *and* Y *are* independent *random variables, then* $E(XY) = E(X)E(Y)$.

Proof:

Since X and Y are independent,

$$p(x, y) = p_1(x)p_2(y)$$

for all pairs of x and y. Hence,

$$E(XY) = \sum_{x}\sum_{y} xyp(x, y).$$

Substituting for p(x, y) gives

$$E(XY) = [\sum_{y} yp_2(y)][\sum_{x} xp_1(x)].$$

Using the definition of expectation, we have

$$E(XY) = E(X)E(Y). \qquad \square$$

As a consequence of Theorem 8·1, we see that when X and Y are independent random variables

$$\text{Cov}(X, Y) = 0.$$

We found that in two of our examples, the covariance was zero. An examination of the joint probability functions shows that only in Example 3 are X and Y independent. *A zero covariance does not imply independence. If, however, the random variables are independent, then their covariance is zero.*

8-3 Expectation of the Sum of Several Random Variables

At this point it may be useful to review Section 7-2. We will now extend the concepts developed there to the expectation of the sum of several random variables.

> Theorem 8·2: *If X and Y are any two random variables, then* $E(X + Y) = E(X) + E(Y)$.

Proof:

Let X and Y have joint p.f. $p(x, y)$. Then the marginal p.f.'s are $p_1(x) = \sum_y p(x, y)$ and $p_2(y) = \sum_x p(x, y)$. Be Definition 8·1,

$$E(X + Y) = \sum_x\sum_y (x + y)p(x, y)$$

$$= \sum_x\sum_y xp(x, y) + \sum_x\sum_y yp(x, y).$$

Using the definition of marginal p.f.'s, we can rewrite the r.h.s. and obtain

$$E(X + Y) = \sum_x x[\sum_y p(x, y)] + \sum_y y[\sum_x p(x, y)]$$

$$= \sum_x xp_1(x) + \sum_y yp_2(y).$$

Then by the definition of expectation, we have

$$E(X + Y) = E(X) + E(Y). \qquad \square$$

Note that in contrast to Theorem 8·1 this property of expectation does not require the condition of independence. It is true for both dependent and independent random variables.

Theorem 8·2 can easily be extended to k random variables. The proof of the following theorem is left as an exercise.

Theorem 8·3: *If $X_1, X_2, \ldots, X_k$ are k random variables, then the expectation of the linear function $\sum_{i=1}^{k} a_i X_i$, where the a_i's are constants, is*

$$E(\sum_{i=1}^{k} a_i X_i) = \sum_{i=1}^{k} a_i E(X_i).$$

In the special case where $a_i = 1$ for all i, Theorem 8·3 reduces to *the expectation of the sum is equal to the sum to the individual expectations.*

8-4 Variance of the Sum of Several Random Variables

In this section we will extend the results of Section 7-4.

Theorem 8·4: *If X and Y are any two random variables, then*

$$\mathrm{Var}(X \pm Y) = \mathrm{Var}(X) + \mathrm{Var}(Y) \pm 2\,\mathrm{Cov}(X, Y).$$

Proof:

Starting with the definition of variance, we have

$$\text{Var}(X + Y) = E\{(X + Y) - E(X + Y)\}^2.$$

We then regroup the r.h.s.:

$$\text{Var}(X + Y) = E\{[X - E(X)] + [Y - E(Y)]\}^2.$$

Squaring the binomial, we get

$$\text{Var}(X + Y) = E\{[X - E(X)]^2\} + E\{[Y - E(Y)]^2\} + 2E\{[X - E(X)][Y - E(Y)]\}.$$

Applying the definition of variance and covariance to the r.h.s., the variance of the sum becomes

$$\text{Var}(X + Y) = \text{Var}(X) + \text{Var}(Y) + 2\,\text{Cov}(X, Y).$$

The results for the difference (X - Y) follow in a similar manner to give

$$\text{Var}(X - Y) = \text{Var}(X) + \text{Var}(Y) - 2\,\text{Cov}(X, Y).$$ □

How can these results be extended to the weighted sum of k random variables?

Theorem 8·5: *Let* $X_1, X_2, \ldots, X_k$ *be* k *random variables, then the variance of their weighted sum*

$$a_1X_1 + a_2X_2 + \cdots + a_kX_k,$$

where $a_1, a_2, \ldots, a_k$ *are constants, is given by*

$$\text{Var}\left(\sum_{i=1}^{k} a_iX_i\right) = \sum_{i=1}^{k} a_i^2\text{Var}(X_i) + \sum_{i \neq j}\sum a_ia_j\text{Cov}(X_i, X_j) \quad (1)$$

or

$$\text{Var}\left(\sum_{i=1}^{k} a_iX_i\right) = \sum_{i=1}^{k} a_i^2\text{Var}(X_i) + 2\sum_{i<j}\sum a_ia_j\text{Cov}(X_i, X_j). \quad (2)$$

Proof:

This proof is quite similar to that of Theorem 8·4. In the previous theorem, we squared the binomial

$$\{[X - E(X)] + [Y - E(Y)]\}.$$

Now we extend this to the squaring of the multinomial

$$\{a_1[X_1 - E(X_1)] + a_2[X_2 - E(X_2)] + \cdots + a_k[X_k - E(X_k)]\}.$$

Squaring this multinomial, we get

$$\begin{aligned} a_1^2[X_1 - E(X_1)]^2 + a_2^2[X_2 - E(X_2)]^2 + \cdots + a_k^2[X_k - E(X_k)]^2 \\ + a_1a_2[X_1 - E(X_1)][X_2 - E(X_2)] + \cdots \\ + a_1a_k[X_1 - E(X_1)][X_k - E(X_k)] + \cdots \\ + a_ka_1[X_k - E(X_k)][X_1 - E(X_1)] + \cdots \\ + a_ka_{k-1}[X_k - E(X_k)][X_{k-1} - E(X_{k-1})]. \end{aligned}$$

Now apply the expectation operator to each term and note its form. There are k terms, in the sum, of the general form

$$E\{a_i^2[X_i - E(X_i)]^2\} = a_i^2\text{Var}(X_i).$$

Thus the k terms of this form can be summarized as

$$\sum_{i=1}^{k} a_i^2\text{Var}(X_i).$$

Now consider the cross-product terms

$$E\{a_ia_j[X_i - E(X_i)][X_j - E(X_j)]\} = a_ia_j\text{Cov}(X_i, X_j).$$

How many such terms are there? There will be $k(k - 1)$ such terms in the sum. In more compact notation we have

$$\sum_{i \neq j}\sum a_ia_j\text{Cov}(X_i, X_j).$$

Since the $\text{Cov}(X_i, X_j)$ is the same as $\text{Cov}(X_j, X_i)$, pairs of terms in this sum will be the same; hence, we can write this as

$$2 \sum_{i<j}\sum a_i a_j \text{Cov}(X_i, X_j),$$

where this summation is now over k(k - 1)/2 terms. In summary, then, the $\text{Var}(\sum_{i=1}^{k} a_i X_i)$ can be written in two ways:

$$(1) \quad \sum_{i=1}^{k} a_i^2 \text{Var}(X_i) + \sum_{i \neq j}\sum a_i a_j \text{Cov}(X_i, X_j)$$

$$(2) \quad \sum_{i=1}^{k} a_i^2 \text{Var}(X_i) + 2 \sum_{i<j}\sum a_i a_j \text{Cov}(X_i, X_j). \qquad \square$$

Note that if $a_1 = a_2 = \cdots = a_k = 1$, Theorem 8·5 simplifies to

$$\text{Var}(\sum_{i=1}^{k} X_i) = \sum_{i=1}^{k} \text{Var}(X_i) + \sum_{i \neq j}\sum \text{Cov}(X_i, X_j)$$

or

$$\text{Var}(\sum_{i=1}^{k} X_i) = \sum_{i=1}^{k} \text{Var}(X_i) + 2 \sum_{i<j}\sum \text{Cov}(X_i, X_j).$$

If the random variables are <u>independent</u>, the variance of the sum is simplified. How?

Theorem 8·6: *Let* $X_1, X_2, \ldots, X_k$ *be* k <u>independent</u> *random variables. Then*

$$\text{Var}(\sum_{i=1}^{k} a_i X_i) = \sum_{i=1}^{k} a_i^2 \text{Var}(X_i),$$

where $a_1, a_2, \ldots, a_k$ *are constants.*

Proof:

If the X_i's are independent, then

$$\text{Cov}(X_i, X_j) = 0 \qquad \text{for } i \neq j.$$

If the covariances are zero, then the second term of the sum given in Theorem 8·5 disappears, and we have

$$\mathrm{Var}\left(\sum_{i=1}^{k} a_i X_i\right) = \sum_{i=1}^{k} a_i^2 \mathrm{Var}(X_i). \qquad \square$$

Again if the a_i's are all one and the X_i's are independent, then

$$\mathrm{Var}\left(\sum_{i=1}^{k} X_i\right) = \sum_{i=1}^{k} \mathrm{Var}(X_i).$$

The results obtained in the previous theorems may be illustrated more conveniently by using an array or matrix notation. For example, if we consider the simple case of two random variables X_1 and X_2, we can summarize their variance-covariance structure in an array

$$\begin{bmatrix} \mathrm{Var}(X_1) & \mathrm{Cov}(X_1, X_2) \\ \mathrm{Cov}(X_2, X_1) & \mathrm{Var}(X_2) \end{bmatrix}$$

Note that this array is symmetric since $\mathrm{Cov}(X_1, X_2) = \mathrm{Cov}(X_2, X_1)$. How do we use this array to find $\mathrm{Var}(X_1 + X_2)$? Clearly, $\mathrm{Var}(X_1 + X_2)$ = the sum of all the terms in the array. In general for k r.v.'s we would have a $k \times k$ array

$$\begin{bmatrix} \mathrm{Var}(X_1) & \mathrm{Cov}(X_1, X_2) & \cdots & \mathrm{Cov}(X_1, X_k) \\ \mathrm{Cov}(X_1, X_2) & \mathrm{Var}(X_2) & \cdots & \mathrm{Cov}(X_2, X_k) \\ \vdots & & & \vdots \\ \mathrm{Cov}(X_1, X_k) & \mathrm{Cov}(X_2, X_k) & \cdots & \mathrm{Var}(X_k) \end{bmatrix}$$

Again,

$$\mathrm{Var}\left(\sum_{i=1}^{k} X_i\right) = \text{sum of the terms in array.}$$

An examination of this array should easily clarify the number of terms in the second sum of the two expressions for $\mathrm{Var}(\sum_{i=1}^{k} X_i)$ given in Theorem 8·5. It should also demonstrate the equivalence of the two forms.

If the X_i's are independent, then the $\mathrm{Cov}(X_i, X_j)$ will be zero for all values of i and j. Thus all the off-diagonal terms in the array will be zero.

The concepts introduced in Sections 8-3 and 8-4 will be further illustrated by examples in Section 8-6 of this chapter and in the next chapter.

8-5 Correlation Coefficient

We have seen that the covariance can be used as a measure of the dependence between two random variables. We would, however, like any such measure to be free of the units in which X_1 and X_2 are measured. We know that the covariance does not have this property.

Definition 8·3: *Let* X *and* Y *be random variables with expected values* E(X), E(Y) *and nonzero variances* σ_X^2, σ_Y^2. *Then the* correlation coefficient between X and Y *is given by*

$$\rho_{XY} = \mathrm{Corr}(X, Y) = \frac{E\{[X - E(X)][Y - E(Y)]\}}{\sigma_X \sigma_Y} = \frac{\mathrm{Cov}(X, Y)}{\sigma_X \sigma_Y}.$$

From Definition 8·3 and Theorem 8·1 it is clear that ρ_{XY}, like the covariance, will be zero if X and Y are independent. Again we should be fully aware that a zero correlation coefficient does not imply independence.

Suppose we know σ_X^2, σ_Y^2 and the correlation coefficient ρ_{XY}. Is is possible to determine Cov(X, Y)? From Definition 8·3 we have

$$\mathrm{Cov}(X, Y) = \rho_{XY}\sigma_X\sigma_Y.$$

We will find that the definition in this form will often be useful in solving problems in which the variances and correlation coefficient are specified.

We mentioned that the correlation coefficient is a scale-free measure. In the following theorems we will show that

$$-1 \le \rho_{XY} \le 1$$

and that these bounds are attained when Y is a linear function of X.

Theorem 8·7: *The correlation coefficient* ρ_{XY} *between two random variables* X *and* Y *is a number between* -1 *and* +1 *inclusive:*

$$-1 \le \rho_{XY} \le 1.$$

Proof:

As a part of this proof, we will introduce the concept of standardized random variables. Frequently, it may be useful to transform a r.v., say Z, with expectation E(Z) and variance Var(Z) to a new r.v. Z^* with expectation zero and variance 1. What is the form of this transformation? Take

$$Z^* = \frac{Z - E(Z)}{\sqrt{Var(Z)}} .$$

It is quite easy to show that

$$E(Z^*) = 0 \quad \text{and} \quad Var(Z^*) = 1.$$

We will refer to Z^* as a standardized random variable.

Let us now standardize the r.v.'s X and Y:

$$X^* = \frac{X - E(X)}{\sqrt{Var(X)}}$$

$$Y^* = \frac{Y - E(Y)}{\sqrt{Var(Y)}}$$

with $E(X^*) = E(Y^*) = 0$ and $\text{Var}(X^*) = \text{Var}(Y^*) = 1$. Now consider the variance of the sum $(X^* + Y^*)$. By Theorem 8·4,

$$\text{Var}(X^* + Y^*) = \text{Var}(X^*) + \text{Var}(Y^*) + 2\text{Cov}(X^*, Y^*).$$

By definition,

$$\text{Cov}(X^*, Y^*) = E(X^*Y^*) - E(X^*)E(Y^*).$$

Since $E(X^*) = E(Y^*) = 0$,

$$\text{Cov}(X^*, Y^*) = E\left\{\frac{[X - E(Y)][Y - E(Y)]}{\sqrt{\text{Var}(X)\text{Var}(Y)}}\right\}$$

$$= \rho_{XY}.$$

Recalling that $\text{Var}(X^*) = \text{Var}(Y^*) = 1$, we have

$$\text{Var}(X^* + Y^*) = 2(1 + \rho_{XY}).$$

We know that the variance of any r.v. must be greater than or equal to zero; thus, it follows that

$$2(1 + \rho_{XY}) \geq 0$$

which implies that

$$\rho_{XY} \geq -1.$$

Similarly,

$$\text{Var}(X^* - Y^*) = 2(1 - \rho_{XY}),$$

which implies that

$$\rho_{XY} \leq 1.$$

Therefore, combining these results, we have

$$-1 \leq \rho_{XY} \leq 1. \qquad \square$$

Theorem 8·8: *If Y is a linear function of X, that is* $Y = aX + b$, *then* ρ_{XY} *is +1 if a is positive and -1 if a is negative.*

Proof:

Since $Y = aX + b$, $E(Y) = aE(X) + b$; therefore,

$$Y - E(Y) = a[X - E(X)].$$

Using Theorem 7·8 and Definition 8·2,

$$\text{Var}(Y) = a^2\text{Var}(X)$$

and

$$\text{Cov}(X, Y) = a\text{Var}(X);$$

hence,

$$\rho_{XY} = \frac{a\text{Var}(X)}{\sqrt{a^2\text{Var}(X)\text{Var}(X)}} = \frac{a}{|a|}.$$

If a is positive, ρ_{XY} is +1 and if a is negative, ρ_{XY} is -1. □

8-6 Problems Concerning Several Random Variables

The techniques of this chapter may often be useful in solving problems which do not seem to involve several variables. We shall now examine, in detail, several examples.

Example 4: Let us again return to our experiment of tossing two coins: a nickel and a penny. Suppose the person gets to keep the coin on which a head appears. Let his reward be symbolized by the r.v. R. Note that R has a value set {0, 1, 5, 6}. Here we can decompose R into the sum of two random variables:

X_1 = reward if nickel shows a head

X_2 = reward if penny shows a head.

Then

$$R = X_1 + X_2.$$

Find the expectation and variance of R.

Solution: $E(R) = E(X_1) + E(X_2) = 5/2 + 1/2 = 3$. Since X_1 and X_2 are independent,

$$\text{Var}(R) = \text{Var}(X_1) + \text{Var}(X_2) = \frac{25}{4} + \frac{1}{4} = \frac{26}{4} .$$

Example 5: Consider two indicator random variables: I_A as an indicator of the event A and I_B as an indicator of the event B. Find the $\text{Cov}(I_A, I_B)$.

Solution: Using the computational formula, we have

$$\text{Cov}(I_A, I_B) = E(I_A I_B) - E(I_A)E(I_B).$$

We know that

$$E(I_A) = \Pr(A)$$

and

$$E(I_B) = \Pr(B).$$

Recall that the product $I_A I_B$ is an indicator of the event $A \cap B$; hence,

$$E(I_A I_B) = \Pr(A \cap B).$$

Combining these results, we get

$$\text{Cov}(I_A I_B) = \Pr(A \cap B) - \Pr(A)\Pr(B).$$

In passing, we might point out that if the events A and B are independent, then their indicators I_A and I_B will also be independent; hence, $\text{Cov}(I_A, I_B) = 0$. In the particular case of indicator random variables, $\text{Cov}(I_A, I_B) = 0$ implies that the indicators I_A and I_B as well as the events A and B are independent. Proof of this converse statement is left as an exercise.

Example 6: We will now return to the "matching problem." Suppose we have a deck of N cards labeled 1, 2, ..., N. We deal the cards one at a time. If the card numbered i is dealt on the ith deal, we say that a match has occurred. Determine the mean and variance of the total number of matches.

Solution: Let T be a random variable counting the total number of matches. Clearly a match can occur on any of the N deals. If X_i is an indicator for a match on the ith deal, then

$$T = X_1 + X_2 + \cdots + X_N.$$

What is the p.f. of X_i? By the equivalence law the probability of a match on any deal is 1/N; hence, the p.f. of X_i is

x_i	0	1
$f(x_i)$	$1 - \frac{1}{N}$	$\frac{1}{N}$

with

$$E(X_i) = \frac{1}{N} \quad \text{and} \quad Var(X_i) = \frac{1}{N}(1 - \frac{1}{N}).$$

Now applying Theorem 8·2, we can find the E(T):

$$E(T) = E(\sum_{i=1}^{N} X_i) = \sum_{i=1}^{N} E(X_i) = \sum_{i=1}^{N} \frac{1}{N} = 1.$$

Now if we use Theorem 8·5, we can write

$$Var(T) = Var(\sum_{i=1}^{N} X_i) = \sum_{i=1}^{N} Var(X_i) + 2 \sum_{i<j}\sum Cov(X_i, X_j).$$

Since X_i and X_j are indicators, we know from Example 5 that

$$\begin{aligned} Cov(X_i, X_j) &= Pr(\text{match on ith and jth deal}) \\ &\quad - Pr(\text{match on ith deal})Pr(\text{match on jth deal}) \\ &= Pr(X_i = 1 \text{ and } X_j = 1) - Pr(X_i = 1)Pr(X_j = 1). \end{aligned}$$

Let us now consider the probabilities of these three events. If the ith and jth cards are fixed in their respective places, then the remaining N - 2 cards can be dealt at random:

$$\Pr(X_i = 1 \text{ and } X_j = 1) = \frac{(N-2)!}{N!} = \frac{1}{N(N-1)} .$$

From the p.f. of X_i, we know that

$$\Pr(X_i = 1) = \Pr(X_j = 1) = \frac{1}{N} ;$$

therefore,

$$\text{Cov}(X_i, X_j) = \frac{1}{N(N-1)} - \frac{1}{N^2} .$$

Substituting for $\text{Var}(X_i)$ and $\text{Cov}(X_i, X_j)$ in the formula for Var(T), we get

$$\text{Var}(T) = \sum_{i=1}^{N} \frac{1}{N}\left(1 - \frac{1}{N}\right) + 2 \sum_{i<j}\sum \left[\frac{1}{N(N-1)} - \frac{1}{N^2}\right].$$

Note that for both the first and second terms, the expression following the summation symbol is a constant. The first summation has N terms and the double summation has $N(N-1)/2$ terms; hence,

$$\text{Var}(T) = 1.$$

Recall that when we discussed the matching problem for $N = 3$, we did find that $E(T) = 1$ and $\text{Var}(T) = 1$. This example, however, shows that this result holds regardless of the number of cards.

Example 7: Suppose Jim and John decide to toss a balanced coin. Jim says he wants to count the number of heads X while John prefers to count the number of tails Y. Find (a) Cov(X, Y), (b) ρ_{XY}. (c) Are X and Y independent?

Solution: (a) Here again we are dealing with indicator random variables:

$$E(X) = E(Y) = \frac{1}{2}$$

$$\text{Var}(X) = \text{Var}(Y) = \frac{1}{4} .$$

As before, we can write

$$\text{Cov}(X, Y) = \Pr(X = 1 \text{ and } Y = 1) - \Pr(X = 1)\Pr(Y = 1).$$

Now the event $(X = 1 \text{ and } Y = 1)$ is impossible; therefore,

$$\text{Cov}(X, Y) = 0 - \left(\frac{1}{2}\right)\left(\frac{1}{2}\right) = -\frac{1}{4}.$$

(b) By definition,

$$\rho_{XY} = \frac{\text{Cov}(X, Y)}{\sigma_X \sigma_Y} = \frac{-\frac{1}{4}}{\sqrt{(1/4)(1/4)}} = -1.$$

In Theorem 8.8, we saw that if Y is a linear function of X, then $\rho = -1$ or $+1$. Here Y is a linear function of X,

$$Y = N - X.$$

From Theorem 8.8 it follows that $\rho_{XY} = -1$.

(c) Since $\text{Cov}(X, Y) \neq 0$, X and Y are not independent.

Example 8: If X has p.f.

x	-1	0	1
f(x)	.25	.5	.25

(a) Determine Cov(X, Y) when $Y = X^2$. (b) Are X and Y independent?

Solution: We know that $E(X) = 0$, $E(Y) = .5$. Now, to determine Cov(X, Y) we need the joint p.f. of X and Y:

y \ x	-1	0	1
0	0	$\frac{1}{2}$	0
1	$\frac{1}{4}$	0	$\frac{1}{4}$

Using this, then

$$\begin{aligned}\text{Cov}(X, Y) &= E(XY) - E(X)E(Y) \\ &= 0 - 0 = 0.\end{aligned}$$

Since $\text{Cov}(X, Y) = 0$, $\rho_{XY} = 0$.

(b) We have found that $\rho_{XY} = 0$; hence, X and Y may or may not be independent. To verify that X and Y are independent, we must check that

$$f(x, y) = f_1(x)f_2(y)$$

for all x and y. Since this condition does not hold, X and Y are <u>not</u> independent.

This example points out that ρ_{XY} is measuring only the strength of a linear relation between X and Y. In this case we had a parabolic relation: $Y = X^2$; however, $\rho = 0$--indicating no linear relation.

Example 9: Let $X_1, X_2, \ldots, X_k$ be a set of random variables, each with expectation τ and variance η. The covariances are

$$\text{Cov}(X_i, X_{i+1}) = \rho\eta$$

for $i = 1, 2, \ldots, (k - 1)$, and all other covariances are zero. This form of dependence is a particular case of serial correlation. Find the expected value and variance of $\frac{1}{k}\sum_{i=1}^{k} X_i$.

Solution: The expected value is not affected by dependence; hence, $E(\frac{1}{k}\sum_{i=1}^{k} X_i) = \tau$. The variance is

$$\text{Var}\left(\frac{1}{k}\sum_{i=1}^{k} X_i\right) = \frac{1}{k^2}\left[\sum_{i=1}^{k} \text{Var}(X_i) + \sum_{i\neq j}\sum \text{Cov}(X_i, X_j)\right].$$

Let us consider the term involving the covariance. To visualize how many covariances are nonzero, let us restrict ourselves to k = 4 and then form the array of the variance-covariance structure:

$$\begin{bmatrix} \eta & \rho\eta & 0 & 0 \\ \rho\eta & \eta & \rho\eta & 0 \\ 0 & \rho\eta & \eta & \rho\eta \\ 0 & 0 & \rho\eta & \eta \end{bmatrix}$$

Note that a diagonal of nonzero terms forms above and below the main diagonal. If there are k r.v.'s, how many such terms will there be? These diagonals each contain k - 1 terms; thus, the summation $\sum_{i \neq j}\sum$ has 2(k - 1) nonzero terms. In summary,

$$\mathrm{Var}\left(\frac{1}{k}\sum_{i=1}^{k} X_i\right) = \frac{1}{k^2} k\eta + \frac{2(k-1)}{k^2}\rho\eta$$

$$= \frac{\eta}{k}\left\{1 + 2\rho\left(1 - \frac{1}{k}\right)\right\}.$$

8-7 Random Variables Based on Samples

In Chapter 1 we discussed the role of statistics in making inferences about populations based on samples. We will now show how theorems for the mean and variance of sums of independent random variables can be applied to sampling.

Suppose a box contains N balls each having a number on it. The same number may appear on several balls. Let $y_1, y_2, \ldots, y_k$ for $k \leq N$ be the distinct numbers on the balls. If n_j is the number of balls with the number y_j, then the relative frequency or proportion of balls having the number y_j is n_j/N. Note that

$$\sum_{j=1}^{k} n_j = N \quad \text{and} \quad \sum_{j=1}^{k} \frac{n_j}{N} = 1.$$

If one ball is selected at random and we let Y be the number on the ball, then the probability function of Y is

y	y_1	y_2	...	y_k
Pr(Y = y)	$\frac{n_1}{N}$	$\frac{n_2}{N}$	...	$\frac{n_k}{N}$

or

$$f(y_j) = \frac{n_j}{N} \quad \text{for } j = 1, 2, \ldots, k.$$

We can easily see that the mean and variance of Y are given by

$$\mu_Y = \frac{1}{N} \sum_j y_j n_j$$

$$\sigma_Y^2 = \frac{1}{N} \sum_j (y_j - \mu_Y)^2 n_j .$$

These quantities are referred to as the *population mean* and *population variance* of Y.

To make these ideas somewhat more concrete, we might think of these N balls as a population of N people with the number on the ball being the person's age.

Now suppose we select n balls with replacement. Since the selection is with replacement, each selection then forms an independent trial. We can visualize this as a n-stage experiment and the assignment of probabilities is made using the product rule for independent events.

Let Y_i be a random variable whose value is the number on the ith ball drawn for i = 1, 2, ..., n. Thus for the n-stage experiment we have n independent random variables $Y_1, Y_2, \ldots, Y_n$, each of which has the same probability function. This set of random variables is called a random sample from the population of Y. We say that the random variables $Y_1, Y_2, \ldots, Y_n$ are independent and identically distributed random variables with mean μ_Y and variance σ_Y^2.

How can we combine the information obtained from the n draws? Frequently we may summarize the sample information by using the sample mean

$$\bar{Y} = \frac{Y_1 + Y_2 + \cdots + Y_n}{n} .$$

Since $\bar{Y}$ is a function of random variables, it too is a random variable.

If we know the probability function of Y_i, what can we say about the probability function of $\bar{Y}$? Does a knowledge of μ_Y and σ_Y^2 provide any information about $\mu_{\bar{Y}}$ and $\sigma_{\bar{Y}}^2$?

To investigate these questions consider a random variable X with p.f.

x	-1	0	2
f(x)	.3	.5	.2

Consider all the possible samples of size 2 that could be drawn with replacement from this population:

Sample	*Probability of Sample*	*Value of* $\bar{X}$
(-1, -1)	.09	-1
(-1, 0)	.15	-.5
(-1, 2)	.06	.5
(0, -1)	.15	-.5
(0, 0)	.25	0
(0, 2)	.10	1
(2, -1)	.06	.5
(2, 0)	.10	1
(2, 2)	.04	2

Using the above table we can determine the p.f. of $\bar{X}$, the sample mean:

$\bar{x}$	-1	-.5	0	.5	1	2
$\Pr(\bar{X} = \bar{x})$	.09	.30	.25	.12	.20	.04

It can be easily verified from p.f. of X and of $\bar{X}$ that

$$\mu_X = .1 \quad \text{and} \quad \sigma_X^2 = 1.09$$

$$\mu_{\bar{X}} = .1 \quad \text{and} \quad \sigma_{\bar{X}}^2 = .545.$$

Note that

$$\mu_{\bar{X}} = \mu_X \quad \text{and} \quad \sigma_{\bar{X}}^2 = \frac{\sigma_X^2}{2}.$$

With these results at hand, we will prove the following theorem:

> Theorem 8·9: *Let* $X_1, X_2, \ldots, X_n$ *be* n *independent, identically distributed random variables, each with mean* μ_X *and variance* σ_X^2. *If*
>
> $$\bar{X} = \frac{1}{n} \sum_{i=1}^{n} X_i,$$
>
> *then*
>
> $$\mu_{\bar{X}} = \mu_X \qquad \text{and} \qquad \sigma_{\bar{X}}^2 = \frac{\sigma_X^2}{n}.$$

Proof:

$$\mu_{\bar{X}} = E(\bar{X}) = E\left(\frac{1}{n} \sum_{i=1}^{n} X_i\right) = \frac{1}{n} \sum_{i=1}^{n} E(X_i)$$

$$= \frac{1}{n} \sum_{i=1}^{n} \mu_X = \frac{1}{n} (n\mu_X) = \mu_X$$

$$\sigma_{\bar{X}}^2 = \text{Var}(\bar{X}) = \text{Var}\left(\frac{1}{n} \sum_{i=1}^{n} X_i\right)$$

$$= \frac{1}{n^2} \sum_{i=1}^{n} \text{Var}(X_i) = \frac{1}{n^2} \sum_{i=1}^{n} \sigma_X^2$$

$$= \frac{1}{n^2} (n\sigma_X^2) = \frac{\sigma_X^2}{n}.$$

□

What are the implications of this theorem? The expected value of the sample mean $\bar{X}$ is the same as the population mean μ. The variance of the sample mean is reduced by a factor of $1/n$. Thus, as the sample size n increases, the variance of the distribution of $\bar{X}$ is reduced and the values of the sample mean become more concentrated about the population mean μ.

How can we ensure that the sample mean $\bar{X}$ is near the population mean μ? Theorem 8·10, which is known as the *law of large numbers*, tells us that by choosing the sample size n sufficiently large, the probability that the value of the sample mean $\bar{X}$ differs from the population mean by at most C can be made as close to 1 as we like.

Theorem 8·10: *Let X be a random variable with mean μ_X and variance σ_X^2. If $\bar{X}$ is the mean of a random sample of size n drawn with replacement from this population and C is any positive number, then as n increases without bound*

$$\Pr(\mu_X - C \le \bar{X} \le \mu_X + C)$$

approaches 1.

Proof:

Applying Chebshev's inequality to the random variable $\bar{X}$, we find that

$$\Pr\{|\bar{X} - \mu_{\bar{X}}| > C\} < \frac{\sigma_{\bar{X}}^2}{C^2}.$$

Writing $\mu_{\bar{X}}$ and $\sigma_{\bar{X}}^2$ in terms of μ_X and σ_X^2, we have

$$\Pr\{|\bar{X} - \mu_X| > C\} < \frac{\sigma_X^2}{nC^2}$$

or

$$\Pr\{|\bar{X} - \mu_X| \le C\} > 1 - \frac{\sigma_X^2}{nC^2}.$$

As n increases, the quantity σ_X^2/nC^2 approaches zero and

$$1 - \frac{\sigma_X^2}{nC^2}$$

approaches 1. Thus by choosing n sufficiently large

$$\Pr\{|\bar{X} - \mu_X| \le c\}$$

approaches 1. □

Problems

1. Prove that

$$\text{Var}(X - Y) = \text{Var}(X) + \text{Var}(Y) - 2\,\text{Cov}(X, Y).$$

2. In three tosses of a balanced coin, let

$$Z = \text{number of tails}$$
$$W = \text{number of heads.}$$

   (a) Find E(Z), E(W), Var(Z), Var(W).

   (b) Construct the joint p.f. of Z and W. Find Cov(Z, W). Are Z and W independent?

   (c) Find the p.f. of Z + W. From this p.f., determine E(Z + W) and Var(Z + W).

   (d) Verify that the results in (c) are the same as those obtained by using Theorems 8·2 and 8·4.

3. A die is loaded so that the probability associated with a given face is proportional to the number on that face. Suppose such a die is rolled once. Let

   X = twice the number appearing on it

   Y = 1 if an odd number appears and 3 if an even number appears.

   (a) Find the probability functions of X, Y and (X, Y).

   (b) Determine E(XY), E(X) and E(Y).

   (c) Are X and Y independent?

4. Let X and Y have the following joint probability function

x \ y	1	2	3
1	.1	.1	0
2	.1	.2	.3
3	.1	.1	0

(a) Find the probability functions for X + Y, XY, X/Y.

(b) Using the p.f.'s obtained in (a), find E(X + Y), E(XY), E(X/Y), Var(X + Y).

(c) Find the p.f.'s of X and of Y. Determine E(X), E(Y), Var(X) and Var(Y).

(d) Using the results of (b) and (c), compare

(i) E(X + Y) and E(X) + E(Y)

(ii) E(XY) and E(X)E(Y)

(iii) E(X/Y) and E(X)/E(Y).

(e) Are X and Y independent random variables? Compare Var(X + Y) and Var(X) + Var(Y). Comment on this result.

5. The joint p.f. of (X, Y) is given by

x \ y	0	1	2
0	$\frac{1}{6}$	$\frac{1}{6}$	$\frac{1}{6}$
1	$\frac{1}{6}$	$\frac{1}{6}$	0
2	$\frac{1}{6}$	0	0

(a) Find Var(X), Var(Y) and ρ_{XY}.

(b) Determine Var(X + Y).

(c) Are X and Y independent?

(d) Comment on the relation between Var(X) + Var(Y) and Var(X + Y).

6. In a freshman biology course students were given two term tests with the following results

	μ	σ	$Cov(X_1, X_2)$
Test I, X_1	60	10	50
Test II, X_2	70	15	

The instructor is considering two possible weighting schemes:

$$W_1 = \frac{1}{2}X_1 + \frac{1}{2}X_2$$

$$W_2 = \frac{1}{3}X_1 + \frac{2}{3}X_2.$$

(a) Find the mean and variance for each weighting scheme.

(b) Which scheme would you prefer?

7. Repeat Problem 6 if the covariance is -100. How would you interpret this negative covariance? What is its affect on the weighting schemes?

8. Repeat Problem 6 if the covariance is zero. What is the implication of a zero covariance?

9. Let $X_1, X_2, \ldots, X_k$ be k random variables. Show, using <u>mathematical induction</u> and Theorem 8·2, that

$$E(\sum_{i=1}^{k} X_i) = \sum_{i=1}^{k} E(X_i).$$

10. Let X be a random variable with mean μ_X and variance σ_X^2. If a sample of n is taken with replacement, find the mean and variance for the sample total $T = \sum_{i=1}^{n} X_i$.

11. Consider two independent random variables X and Y with the following means and standard deviations:

$$\mu_X = 50; \qquad \sigma_X = 10$$

$$\mu_Y = 60; \qquad \sigma_Y = 15.$$

(a) Find $E(X + Y)$, $Var(X + Y)$, $E(X - Y)$, $Var(X - Y)$.

(b) If X^* and Y^* are the standardized r.v.'s corresponding to the r.v.'s X and Y, respectively, determine $E(X^* + Y^*)$, $E(X^* - Y^*)$, $Var(X^* + Y^*)$, $Var(X^* - Y^*)$.

12. If a and b are nonzero constants show that

(a) $Cov(aX, bY) = abCov(X, Y)$.

(b) $\rho(aX, bY) = \frac{ab}{|ab|}\rho(X, Y)$.

13. If X and Y are random variables each of which takes on only two values, prove that if X and Y are uncorrelated they are also independent.

14. Review the experimental situation of Problem 12 in Chapter 5. Define two random variables

$$X = \begin{cases} 0 & \text{if drug is effective} \\ 1 & \text{if drug is not effective} \end{cases}$$

$$Y = \begin{cases} 0 & \text{if male} \\ 1 & \text{if female} \end{cases}$$

(a) Determine μ_X, σ_X^2, μ_Y, σ_Y^2.

(b) Show that $Cov(X, Y) = (ad - bc)/n^2$.

(c) Recall that we proved in Chapter 5 that the events E and M were independent when $(ad - bc)$ is zero. Can we now say that X and Y are independent if the covariance is zero? Why or why not?

(d) Show that

$$\rho(X, Y) = \frac{ad - bc}{\sqrt{(a + b)(c + d)(a + c)(b + d)}}$$

15. Let X be an indicator for the event A and Y an indicator for the event B. Find $\rho(X, Y)$ and hence determine whether or not X and Y are independent when

    (a) $Pr(A) = 1/4$, $Pr(A \mid B) = 1/4$, $Pr(B \mid A) = 1/2$

    (b) $Pr(A) = 1/4$, $Pr(A \mid B) = 3/4$, $Pr(B \mid A) = 1/2$.

16. In Problem 15 of Chapter 6, find $\rho(X, Y)$.

17. If X and Y are <u>independent</u> random variables with joint p.f. $f(x, y)$ and marginal p.f.'s $f_1(x)$ and $f_2(y)$, respectively, prove, using the definition of variance, that

$$Var(X + Y) = Var(X) + Var(Y).$$

18. Let $X_1, X_2, \ldots, X_k$ be k mutually independent random variables. Show, <u>using mathematical induction</u>, that

    (a) $E\left(\prod_{i=1}^{k} X_i\right) = \prod_{i=1}^{k} E(X_i)$

    (b) $Var\left(\sum_{i=1}^{k} X_i\right) = \sum_{i=1}^{k} Var(X_i)$.

    [Hint: In part (a) use Theorem 8·1 and in part (b) use problem 17.]

19. If X and Y are independent random variables, show that

$$E(X_1^2 X_2^2) = E(X_1^2)E(X_2^2).$$

    [Hint: Use Definition 8·1.]

20. Let X have a standard deviation of σ and Y a standard deviation of $k\sigma$ where k is much greater than one. If X and Y are independent, find the approximate standard deviation of $U = X + Y$. Comment on the practical implications of this result.

21. Show that in terms of a standardized random variable X^*, Chebyshev's inequality can be written as

$$Pr(|X^*| \leq h) > 1 - \frac{1}{h^2}$$

    for any positive constant h.

22. Let X^* be a standardized random variable. Using Chebyshev's inequality, determine
    (a) $\Pr(|X^*| > 2)$
    (b) $\Pr(|X^*| \le 2)$.

    What value of k will guarantee that

$$\Pr(|X^*| \le k) > .98?$$

23. Consider n random variables each with variance σ^2 and each pair with correlation ρ.
    (a) Find the variance of the sum $S = X_1 + X_2 + \cdots + X_n$.
    (b) If you are told that the variance of S is zero and $\sigma^2 \ne 0$, what can you say about ρ?

24. Let X_1, X_2, X_3 be three independent random variables. Now consider two new functions

$$V = X_1 + X_2$$

$$W = X_1 + X_3.$$

    (a) Show that $\text{Cov}(V, W) = \text{Var}(X_1)$.
    (b) If $\text{Var}(X_1) = \text{Var}(X_2) = \text{Var}(X_3) = \sigma^2$, find $\rho(V, W)$.

25. Two balanced dice are thrown. Let X be the outcome on the first die and Y the outcome on the second. If $Z = X + Y$, find the correlation between X and Z.

26. Three students (A, B and C) are doing a multiple choice test which consists of 10 questions, each of which has five choices. Let X denote the score of A who is guessing; let Y denote the score of B who knows the first five questions and is guessing the second five questions; let Z denote the score of C who copies A's first five answers and guesses his own answers for the remaining five.
    (a) Find expectation and variance of X, Y and Z.
    (b) Which of the pairs (X, Y), (X, Z), (Y, Z) are independent random variables?
    (c) Find the expectation and variance of $X + Y + Z$.

27. A circular dial has five equal sectors: two sectors are marked with a 2, two are marked with a 3, and one is marked with a 4. A pointer mounted on the dial is spun and the number X at which it stops is recorded. Suppose the dial is spun n times and $\bar{X}$ is the sample mean for the n spins. Find $E(\bar{X})$ and $Var(\bar{X})$.

28. A population is specified by the p.f.

x	-1	0	1
f(x)	$\frac{3}{8}$	$\frac{1}{2}$	$\frac{1}{8}$

(a) Find μ_X and σ_X.

(b) List all the possible samples of size 2 which can be taken with replacement. From this list determine the p.f. of the sample mean $\bar{X} = (X_1 + X_2)/2$.

(c) Find $\mu_{\bar{X}}$ and $\sigma_{\bar{X}}$.

29. A study is being conducted to compare the salaries of female and male employees in a government agency. Denote the mean and standard deviation of the female and male salaries by μ_F, σ_F and μ_M, σ_M, respectively. To obtain information independent random samples (WR) of n_F and n_M individuals are taken from the respective groups. For each group the sample mean is computed.

(a) Show that

$$E(\bar{X}_F - \bar{X}_M) = \mu_F - \mu_M$$

$$Var(\bar{X}_F - \bar{X}_M) = \frac{\sigma_F^2}{n_F} + \frac{\sigma_M^2}{n_M}.$$

What are the implications of these results?

(b) If the two population means, μ_F and μ_M, are equal, what is $E(\bar{X}_F - \bar{X}_M)$?

(c) If in addition to the means being equal, the variances are also the same, what is $Var(\bar{X}_F - \bar{X}_M)$?

30. Consider two random independent samples of sizes n_1 and n_2 taken WR from a population with mean μ and variance σ^2. We will use a linear combination of the sample means $\bar{X}_1$ and $\bar{X}_2$ to estimate the population mean:

$$W = c\bar{X}_1 + (1 - c)\bar{X}_2.$$

(a) Find $E(W)$.

(b) Find $Var(W)$.

(c) Determine c so that $Var(W)$ is a minimum. Why is this a useful criterion for choosing c?

(d) If $n_1 = n_2$, what is the value of c that minimizes $Var(W)$?

31. Let X be the number of cars owned by Canadian families. Suppose it is reasonable to assume that the standard deviation of X is .5. How large a sample of families is required to maintain a probability of at least .95 that the mean of the sample will differ from that of the theoretical mean by no more than .4? [Hint: Use Chebyshev's inequality.]

32. Suppose we have addressed 10 envelopes and 10 letters. A practical joker puts the letters into the envelopes at random.

(a) What is the probability that the first letter and envelope are correctly matched? That the jth letter and envelope are correctly matched?

(b) What is the probability that the first letter and envelope <u>and</u> the second letter and envelope are correctly matched? That the ith and the jth letter-envelope combination are correctly matched?

(c) If X_i for $i = 1, 2, \ldots, 10$ is an indicator of a correct letter-envelope match, discuss how $T = \sum_{i=1}^{10} X_i$ will count the total number of letters placed in correct envelopes.

(d) Find $E(T)$.

(e) Find $Var(T)$.

Chapter 9

Special Discrete Probability Models

The discussion of random variables and their properties in the previous chapters has shown that the same distribution may have varied application. In this chapter several useful distributions will be derived in a more general form.

Each section contains (1) an outline of the conditions under which the distribution arises, (2) derivation of its mean and variance and (3) possible applications.

Interrelationships among the distributions will be pointed out by comparing and contrasting their geneses and exploring their limiting properties.

9-1 Binomial Distribution

9-1-1 Genesis of the Distribution: We have already seen that some experiments are composed of a series of independent trials, each with two outcomes. If, in addition, the probability of each outcome remains the same from trial to trial, we will refer to them as *Bernoulli trials*.

A *binomial experiment* then consists of a fixed number, say n, of Bernoulli trials. If we label the two outcomes as "success" (S) or "failure" (F) with Pr(S) = p and Pr(F) = 1 - p = q, then in a binomial experiment we are interested in the probability of exactly x successes in n Bernoulli trials.

As an illustration of a binomial experiment, consider a marksman shooting at a target. Let S be the event that he hits the target and let Pr(S) = p. If he shoots twice and the probability of S remains the same, what is an appropriate probability model for this experiment? We have discussed many examples similar to this one. A sample space consists of the ordered outcomes (S, S), (S, F), (F, S), (F, F). Since the trials are independent and the probabilities are constant from trial to trial, the probabilities associated with each point in S are p^2, pq, qp and q^2, respectively.

Now define a r.v.

X = number of hits in two trials.

Clearly the p.f. of X is

x	0	1	2
f(x)	q^2	2pq	p^2

Note that the probabilities f(x) are the terms in the expansion of $(q + p)^2$ (see Appendix D).

If the marksman shoots three times, the p.f. of X becomes

x	0	1	2	3
f(x)	q^3	$3pq^2$	$3p^2q$	p^3

with f(x) representing the terms of $(q + p)^3$.

Let us now generalize the situation to n shots. What is the p.f. of X? It is clear that X will have the value set {0, 1, ..., n}. To obtain the probabilities associated with each value, note that there are 2^n ordered points in the sample space. How many of these correspond to X = x? Each point in the sample space which consists of a sequence of x S's and (n - x) F's corresponds to X = x. All such points are generated by permuting the letters S and F in

$$\underbrace{S \ . \ . \ . \ S}_{x} \qquad \underbrace{F \ . \ . \ . \ F}_{n - x}$$

We have seen that there are exactly $\frac{n!}{x!(n - x)!} = \binom{n}{x}$ such orderings.

Assuming independent trials and constant probability from trial to trial, each point in the sample space consisting of x S's and (n - x) F's has the probability $p^x q^{n-x}$. Therefore,

$$\Pr(X = x) = \binom{n}{x} p^x q^{n-x}$$

for x = 0, 1, 2, ..., n. Note here that the probabilities are the terms of the expansion of $(q + p)^n$.

Here we have seen that only a knowledge of n, the number of trials, and p, the probability of success, is necessary for finding the probability of x successes in n trials. These constants which specify a probability function or distribution are called *parameters*. We will see that any measure or description of the distribution (e.g., mean, variance, standard deviation) will be a function of its parameters.

Since a distribution is specified by its parameters, we will use the notation b(x; n, p) to designate the binomial distribution with parameters n and p:

$$b(x; n, p) = \binom{n}{x} p^x q^{n-x} \qquad x = 0, 1, \ldots, n;$$

hence, more explicitly the symbol b(x; 10, .3) means

$$\binom{10}{x} (.3)^x (.7)^{10-x} \qquad x = 0, 1, \ldots, 10.$$

We may also have the occasion to use the symbol ~ to mean "is distributed as" or "has the distribution." For example,

$$X \sim b(x; n, p)$$

means X is distributed as a binomial r.v. with parameters n and p.

An examination of the binomial p.f. shows that it would be a tedious task to compute the probabilities associated with it for a large variety of combinations of p and n. Table 1 at the end of the text contains probabilities for commonly used p and n.

To illustrate the use of the tables consider a series of 25 Bernoulli trials with $p = .8$. Suppose we wish to compute $Pr(X = 19)$ and $Pr(X > 19)$. From Table 1 we find

$$Pr(X = 19) = b(19; 25, .8) = .163.$$

To obtain $Pr(X > 19)$, we note that

$$Pr(X > 19) = \sum_{j=20}^{25} b(j; 25, .8) = .618.$$

9-1-2 Mean and Variance: In Chapter 7 we saw that it is often useful to describe a random variable by its mean and variance. Since the binomial probability function depends on only two parameters n and p, it should be possible to obtain the mean and variance of a binomial random variable as a function of n and p.

Theorem 9·1: *If* $X \sim b(x; n, p)$, $E(X) = np$.

Proof (i):

If $X \sim b(x; n, p)$, then the p.f. is

$$f(x) = \binom{n}{x} p^x q^{n-x} \qquad x = 0, 1, \ldots, n; \; 0 \leq p \leq 1; \; q = 1 - p.$$

By definition,

$$E(X) = \sum_{x=0}^{n} xf(x) = \sum_{x=0}^{n} x \binom{n}{x} p^x q^{n-x}. \tag{1}$$

Expressing the r.h.s. in terms of factorials, we have

$$E(X) = \sum_{x=0}^{n} x \frac{n!}{x!(n - x)!} p^x q^{n-x}.$$

Since there is no contribution to the sum for $x = 0$ and $x/x! = 1/(x - 1)!$, the expected value becomes

$$E(X) = \sum_{x=1}^{n} \frac{n!}{(x-1)!(n-x)!} p^x q^{n-x}. \tag{2}$$

The remainder of this proof rests on Theorem D·1 of Appendix D, namely

$$(a + b)^t = \sum_{j=0}^{t} \binom{t}{j} a^{t-j} b^j.$$

Motivated by the fact that this identity has a summation index beginning at zero, let us make an adjustment in the index of summation in (2) by taking

$$x - 1 = k.$$

Substituting this change in (2) gives

$$E(X) = \sum_{k=0}^{n-1} \frac{n!}{k!(n-1-k)!} p^{k+1} q^{n-1-k}. \tag{3}$$

In an attempt to cast the r.h.s. of (3) into a form similar to the general expansion, we write

$$E(X) = np\left[\sum_{k=0}^{n-1} \frac{(n-1)!}{k!(n-1-k)!} p^k q^{n-1-k}\right].$$

Now the term in the square brackets is $(q + p)^{n-1}$, which is one since $q + p = 1$. Thus

$$E(X) = np.$$

Proof (ii):

This alternative proof is more elegant and the usefulness of indicator random variables will be quite evident. Let I_j be an indicator of the event, success on the jth trial, with $\Pr(I_j = 1) = p$ for $j = 1, 2, \ldots, n$. Then the binomial r.v. X can be decomposed as

$$X = I_1 + I_2 + \cdots + I_n.$$

Since I_j is an indicator, $E(I_j) = p$. From Theorem 8·3,

$$E(X) = E\left[\sum_{j=1}^{n} (I_j)\right] = \sum_{j=1}^{n} E(I_j) = \sum_{j=1}^{n} p = np. \qquad \square$$

From the computing formula, we know that the variance of a random variable is

$$\mathrm{Var}(X) = E(X^2) - [E(X)]^2.$$

In dealing with probability functions which involve factorials, it is often more convenient to compute the variance in the following way: Write

$$X^2 = X(X - 1) + X.$$

Taking the expectation, we get

$$E(X^2) = E[X(X - 1)] + E(X).$$

Substituting this in Var(X) gives

$$\mathrm{Var}(X) = E[X(X - 1)] + E(X) - [E(X)]^2. \tag{4}$$

> Theorem 9·2: *If* $X \sim b(x; n, p)$, $\mathrm{Var}(X) = npq$.

Proof (i):

Using (4), we see that Var(X) depends on E(X) and E[X(X - 1)]. Since we know that E(X) = np by Theorem 9·1, we need only determine E[X(X - 1)]. Following the techniques employed in proof (i) of Theorem 9·1, it can be shown that

$$E[X(X - 1)] = n(n - 1)p^2. \tag{5}$$

Substituting (5) into (4) gives

$$\mathrm{Var}(X) = npq.$$

Proof (ii):

Again write X as the sum of indicators I_j for $j = 1, 2, \ldots, n$. Since the trials are independent, the I_j's will also be independent. We have shown that

$$\text{Var}(I_j) = pq \qquad \text{for } j = 1, 2, \ldots, n.$$

By Theorem 8.6,

$$\text{Var}(X) = \text{Var}\left(\sum_{j=1}^{n} I_j\right) = \sum_{j=1}^{n} \text{Var}(I_j) = \sum_{j=1}^{n} pq = npq. \qquad \square$$

9-1-3: Applications: Before applying the binomial distribution, we must always ensure that the following conditions obtain:

(i) Fixed number of trials.

(ii) Two outcomes for each trial.

(iii) Probability of each outcome is the same for all trials.

(iv) Independent trials.

Note that the last three conditions define Bernoulli trials.

Example 1: An urn contains a red balls and b blue ones. A ball is drawn at random. Its color is noted and then it is replaced. This process is repeated n times. Find the probability function of Y, the number of red balls drawn.

Solution: Considering the section heading our first reaction would be to say Y is a binomial random variable. But to be on the safe side, let's verify the conditions:

(i) Number of trials (draws) fixed in advance at n.

(ii) Two outcomes at each draw: red or blue.

(iii) $\Pr(\text{red}) = \dfrac{a}{a + b}$ and $\Pr(\text{blue}) = \dfrac{b}{a + b}$ for all trials.

(iv) Trials are independent since sampling is with replacement.

Hence,

$$f(y) = b\left(y;\, n, \frac{a}{a + b}\right) = \binom{n}{y}\left(\frac{a}{a + b}\right)^{y}\left(\frac{b}{a + b}\right)^{n-y} \qquad y = 0, 1, 2, \ldots, n.$$

Note that if the sampling had not been with replacement, the trials would not have been independent and Y would not be a binomial variable.

Example 2: Joe claims that he can call the toss of a coin correctly more often than not. If he is successful in 13 out of 15 tosses, would you accept his claim?

Solution: It is easily verified that X, the number of correct calls, is $b(x; 15, p)$. Note that we have not specified p. If Joe is just guessing, then $p = .50$, but if he has some "power" to call the toss, then p would be greater than .50. Assuming Joe is just guessing, $X \sim b(x; 15, .50)$ and from Table 1 we can compute

$$\Pr(X \geq 13) = 1 - \Pr(X \leq 12) = .004.$$

If Joe has power, p will be greater than .5. Let us now look at the same probability, $\Pr(X \geq 13)$, for larger values of p:

p	.6	.8	.95
$\Pr(X \geq 13)$	.027	.398	.964

If we compare these probabilities, we see that this event $X \geq 13$ becomes more likely as Joe's power increases. Since the probability of this event when $p = .50$ is small, we would conclude that either

1. A rare event has occurred, or
2. Joe has "power."

The interpretation of a probability as being small is somewhat of a personal decision. For instance, you may feel that .10 is small, but your neighbor says that it must be less than .010 to be small. To this extent then, the decision is arbitrary.

Example 3: To assess the effectiveness of his product, a manufacturer of weed killer A runs ten controlled experiments. In each experiment there are two component plots of ground; one is chosen at random and treated with the weed killer A, the other is treated with a similar product B, manufactured by another company. Identical crops are grown on each of the two plots and at the end of the experimental period, it is determined whether A outperforms B in each experiment.

(a) Suppose we hypothesize for the moment that in fact A and B are equally effective. Explain why under this hypothesis the distribution of X, the number of experiments in which A outperforms B, is binomial. Specify the parameters of the distribution.

(b) What is $P(X \geq 8)$ under this hypothesis?

(c) Suppose that in eight of the experiments A outperformed B. Is the hypothesis in (a) still tenable?

(d) Consider two alternative experimental situations:

(i) 5 experiments of which A was superior in 4;

(ii) 20 experiments of which A was superior in 16.

Find $P(X \geq 4)$, $P(X \geq 16)$ in these two situations, respectively. How would you answer the second part of question (c) in each of these cases?

(e) Notice that the observed proportion of experiments in which A outperforms B is .8 in all the above cases. What do we gain (and lose) by increasing the number of experiments?

Solution: (a) Checking the conditions for a binomial experiment, we have:

(i) A fixed number of trials: 10 paired plots.

(ii) Two outcomes: A outperforms or B outperforms.

(iii) Pr(A outperforms) is same for all plots.

(iv) The outcomes of the 10 experiments are independent.

It would be realistic to assume that these conditions hold; hence, $X \sim b(x;\ 10,\ p)$. If A and B are equally effective, $p = .5$.

(b) From Table 1 we read

$$\Pr[X \geq 8 \mid b(x;\ 10,\ .5)] = .055.$$

(c) On the basis of our calculation in part (b) we see that the probability seems to be small; hence, we would probably reject the hypothesis and say that A is more effective.

(d) (i) $\Pr[X \geq 4 \mid b(x; 5, .5)] = .187$; the hypothesis is tenable since the probability is reasonably high.

(ii) $\Pr[X \geq 16 \mid b(x; 20, .5)] = .006$; reject the hypothesis since the probability is even smaller than that calculated in (b).

(e) As n increases, we are able to make a more definitive decision based on small probabilities. We do, however, experience a loss in terms of using increased experimentation.

9-1-4 Law of Large Numbers: In the case of binomial random variables the law of large numbers derived by James Bernoulli in the eighteenth century is quite illuminating. Let I_j be an indicator of success on the jth independent trial with $\Pr(I_j = 1) = p$ for $j = 1, 2, \ldots, n$. Then, as we have seen, the random variable $X = \Sigma I_j$ counts the total number of successes in n independent trials:

$$X \sim b(x; n, p).$$

Now consider the proportion of successes in n trials: X/n.

$$E\left(\frac{X}{n}\right) = p$$

$$\mathrm{Var}\left(\frac{X}{n}\right) = \frac{pq}{n}.$$

Applying CHebyshev's inequality for the random variable X/n, we see that

$$\Pr\{\left|\frac{X}{n} - p\right| > c\} < \frac{pq}{nc^2}.$$

Since $pq < 1$, the above inequality can be written as

$$\Pr\{\left|\frac{X}{n} - p\right| > c\} < \frac{1}{nc^2}.$$

Hence the probability that the difference between the proportion of successes and the theoretical probability p is larger than any positive constant c decreases to zero as the number of trials become very large. This provides support for the relative frequency approach to probability. As the number of independent trials is increased, the proportion of successes tends to the true probability of success.

Example 4: A poll is being undertaken to estimate the proportion of people who support the government's new policies on unemployment insurance. How large a sample should be taken so as to ensure with a probability of at least .75 that the estimated proportion differs from the true proportion by no more than .1?

Solution: Using the law of large numbers for proportions, we have

$$\Pr\{\left|\frac{X}{n} - p\right| > c\} < \frac{1}{nc^2}.$$

With c = .1, we wish to determine n such that

$$\Pr\{\left|\frac{X}{n} - p\right| < .1\} \geq .75$$

or

$$\Pr\{\left|\frac{X}{n} - p\right| > .1\} \leq .25.$$

Hence,

$$\frac{1}{n(.1)^2} \leq .25;$$

that is, n must exceed 400. Recall that this form of the law of large numbers was obtained by assuming $pq < 1$. If we use the more restrictive bound

$$pq \leq \frac{1}{4},$$

we can show that

$$\frac{1}{4n(.1)^2} \leq .25$$

or n must exceed 100.

9-2 Waiting Time Distributions

9-2-1 Waiting for the First Success: Again consider, as in Section 9-1, a sequence of Bernoulli trials. Now, however, define a random variable

$$Y = \text{trial number of the first success.}$$

In what way has the experimental set-up changed from that of the binomial experiment? The number of trials is no longer fixed in advance; in fact, the number of trials is a random variable.

If we think about the marksman hitting the target, we are now interested in counting the number of the attempt (trial) on which he first hits the target.

What is the value set for Y in this experiment? Returning to the sample space, we see that there are no longer a finite number of points in it:

$$\text{Sample space} = \{S, FS, FFS, \ldots\}.$$

Clearly the value set of Y, determined on this sample space, will be $1, 2, \ldots, \infty$.

What is the p.f. of Y? Since there is a one-to-one correspondence between the points in the sample space and the values of Y, an assignment of probabilities to the points in the sample space simultaneously gives the p.f. of Y. Since $Pr(S) = p$ and $Pr(F) = q$ for all trials, the probabilities associated with the outcomes are

$$p, qp, q^2p, \ldots, q^rp, \ldots$$

Thus, the p.f. of Y is

$$g(y;\ p) = q^{y-1}p \qquad y = 1, 2, \ldots;\ 0 \le p \le 1;\quad q = 1 - p.$$

We will call this probability function a *geometric distribution* with parameter p.

How is this p.f. different from those we have previously considered? In contrast to previous distributions, the value set of Y is infinite. Is this a proper p.f.? Clearly,

$$0 \le g(y;\ p) \le 1 \qquad \text{for all } y, \text{ since } 0 \le p \le 1.$$

Let us now verify that

$$\sum_{y=1}^{\infty} g(y;\ p) = 1.$$

Consider the l.h.s. in the form

$$\sum_{y=1}^{\infty} q^{y-1}p = p \sum_{y=1}^{\infty} q^{y-1} = p[1 + q + q^2 + \cdots]. \tag{1}$$

Refer to Appendix E and note that the infinite series in (1) is called a geometric series (hence, the name of the distribution). This sum will converge if $|\ q\ | < 1$. Since this condition obtains, the infinite sum converges to

$$(1 - q)^{-1}.$$

Substituting this in (1) and taking $q = 1 - p$, we have

$$\sum_{y=1}^{\infty} q^{y-1}p = p\left[\frac{1}{1 - (1 - p)}\right] = p\left(\frac{1}{p}\right) = 1.$$

Since this is a proper p.f., it implies that no matter how small the probability of success p, a success must occur if the experiment is continued long enough. For example, in genetics, even though a particular mutation is improbable in a given generation, it must eventually appear.

The tail probability, $\Pr(Y > c)$, can be expressed in a form depending only on c and p:

> Theorem 9·3: *If* $Y \sim g(y;\ p)$, *then*
>
> $$\Pr(Y > c) = (1 - p)^c \quad \textit{for } c = 0, 1, 2, \ldots$$

Proof:

For $c = 0$, the result is obvious. For $c > 0$,

$$\Pr(Y > c) = 1 - \sum_{y=1}^{c} pq^{y-1} = 1 - p\left[1 + q + q^2 + \cdots + q^{c-1}\right].$$

The expression in the square brackets is a *finite* geometric series with sum

$$\frac{1 - q^c}{1 - q}.$$

Thus

$$\Pr(Y > c) = 1 - p\left[\frac{1 - (1 - p)^c}{1 - (1 - p)}\right] = (1 - p)^c. \qquad \square$$

The converse of Theorem 9·3 is also true; that is, the geometric distribution is characterized by this property:

> Theorem 9·4: *If* $\Pr(Y > c) = (1 - p)^c$ *for* $c = 0, 1, 2, \ldots,$ then
>
> $$Y \sim g(y;\ p).$$

Proof:

If we know the tail probability, $\Pr(Y > c)$, for all c, how can we find the $\Pr(Y = y)$? By definition

$$\Pr(Y = y) = \Pr(Y > y - 1) - \Pr(Y > y).$$

Using the given statement of the theorem

$$\begin{aligned}\Pr(Y = y) &= (1 - p)^{y-1} - (1 - p)^y \\ &= (1 - p)^{y-1}[1 - (1 - p)] \\ &= (1 - p)^{y-1}p. \qquad \square\end{aligned}$$

Using the results of Appendix E-3, we will find the mean and variance of this distribution.

> Theorem 9·5: *If* $Y \sim g(y; p)$, *then*
>
> $$E(Y) = \frac{1}{p} \quad \textit{and} \quad Var(Y) = \frac{q}{p^2}.$$

Proof:

By definition

$$E(Y) = \sum_{y=1}^{\infty} ypq^{y-1}$$

$$= p[1 + 2q + 3q^2 + \cdots].$$

Since $|q| < 1$, the expression in the square brackets converges to $[1 - q]^{-2}$; hence,

$$E(Y) = p[1 - q]^{-2} = \frac{1}{p}.$$

Recall that

$$Var(Y) = E[Y(Y - 1)] + E(Y) - [E(Y)]^2.$$

Since we already know $E(Y) = 1/p$, we will formally determine $E[Y(Y - 1)]$ as

$$\sum_{y=1}^{\infty} y(y - 1)pq^{y-1}.$$

As there is no contribution to this sum for $y = 1$, we can write

$$E[Y(Y - 1)] = p[2q + (3)2q^2 + (4)3q^3 + \cdots]$$

$$= p(2q)[1 + 3q + 6q^2 + \cdots].$$

Since the sum in the square brackets converges to $[1 - q]^{-3}$,

$$E[Y(Y - 1)] = 2pq[1 - q]^{-3} = \frac{2q}{p^2}.$$

Substituting into the expression for the variance, we have

$$\text{Var}(Y) = \frac{2q}{p^2} + \frac{1}{p} - \frac{1}{p^2} = \frac{q}{p^2}. \qquad \square$$

Example 5: Suppose the relative frequency of a particular blood type is .01. A search is made for a person with this particular blood type by repeated sampling with replacement. (a) What is the expected number of trials necessary to obtain a blood specimen of this type? (b) What is the probability that more than 150 typings are needed before finding the correct type?

Solution: Let X be the trial number on which the correct blood type appears. Then $X \sim g(x; .01)$. (a) The E(X) from Theorem 9·5 is 1/.01 = 100. (b) Using Theorem 9·3, we see that $\Pr(X > 150) = (1 - .01)^{150} = .22$.

9-2-2 Waiting for k Successes: As the title of this section suggests, this distribution arises as a generalization of the geometric distribution. Consider again a sequence of Bernoulli trials in which we wait for k successes. In our problem with the marksman, suppose we wait until he hits the target k times and then we note the number of the trial on which the kth hit occurred. What is the value set of X? In order to have k successes, there must have been at least k trials; hence, the value set of X is k, k + 1, k + 2, ... What is the form of the p.f. of X? A typical point in the sample space consists of x trials with the last trial being a success and (k - 1) successes in the first (x - 1) trials; that is,

$$\underbrace{[k-1]\text{ S's and }[(x-1)-(k-1)]\text{ F's}}_{(x-1)\text{ trials}} \; \Big| \; \underset{x\text{th trial}}{S}$$

The [k - 1] S's and [(x - 1) - (k - 1)] F's in the first (x - 1) trials can be arranged in

$$\frac{(x-1)!}{(k-1)!\,(x-k)!} = \binom{x-1}{k-1}$$

ways. When we include the terminating trial, we see that every point in the sample space consists of k S's and (x - k) F's; hence, the probability of each outcome is $p^k(1 - p)^{x-k}$. The p.f. of X is

$$nb(x;\ p,\ k) = \binom{x-1}{k-1} p^k (1-p)^{x-k} \qquad x = k,\ k+1,\ \ldots,$$

a *negative binomial distribution* with parameters p and **k**.

Again we see that, like the geometric distribution, this distribution takes its name from the negative binomial series. To verify that this is a proper p.f., we must show that

$$\sum_{x=k}^{\infty} \binom{x-1}{k-1} p^k (1-p)^{x-k} = 1.$$

Taking j = x - k and noting that $\binom{j+k-1}{k-1} = \binom{j+k-1}{j}$, the series becomes

$$p^k \left[\sum_{j=0}^{\infty} \binom{j+k-1}{j} (1-p)^j \right].$$

From Appendix E-3 we know that, since $|(1 - p)| < 1$, the series in the square brackets converges to $[1 - (1 - p)]^{-k}$; therefore,

$$\sum_{x=k}^{\infty} \binom{x-1}{k-1} p^k (1-p)^{x-k} = p^k [1 - (1-p)]^{-k} = 1.$$

As might be expected, the mean and variance of this distribution are quite similar to those of the geometric distribution.

> Theorem 9·6: *If* X ~ nb(x; p, k), *then*
>
> $$E(X) = \frac{k}{p} \quad \textit{and} \quad Var(X) = \frac{k(1-p)}{p^2}.$$

Proof:

Like Theorem 9·5, the proof of this theorem depends on the convergence of infinite series. From Appendix E-3 we know that

$$\sum_{j=0}^{\infty} \binom{n + j - 1}{j} t^j = (1 - t)^{-n} \qquad \text{for } |t| < 1.$$

By definition,

$$E(X) = \sum_{x=k}^{\infty} \binom{x - 1}{k - 1} p^k (1 - p)^{x-k} x.$$

Let $j = x - k$ and note that

$$\binom{x - 1}{k - 1} = \binom{x - 1}{x - k} = \binom{x - 1}{j}.$$

Making these substitutions, we have

$$E(X) = \sum_{j=0}^{\infty} \binom{j + k - 1}{j} p^k (1 - p)^j (j + k).$$

Since $(j + k)\binom{j + k - 1}{j} = k\binom{j + k}{j}$,

$$E(X) = \sum_{j=0}^{\infty} p^k k \binom{j + k}{j} (1 - p)^j.$$

Using the general negative binomial expansion with $n = k + 1$ and $t = (1 - p)$,

$$E(X) = p^k k[1 - (1 - p)]^{-(k+1)} = \frac{k}{p}.$$

Using similar techniques, we obtain

$$E[X(X + 1)] = \frac{k(k + 1)}{p^2};$$

hence,

$$\text{Var}(X) = E[X(X + 1)] - E(X) + [E(X)]^2 = \frac{k(1 - p)}{p^2}. \qquad \square$$

Example 6: A balanced coin is tossed repeatedly until three heads are obtained. What is the probability that the third head is obtained before the sixth trial?

Solution: Let X be the trial number. Thus $X \sim nb(x; 1/2, 3)$ and

$$\Pr(X < 6) = \sum_{x=3}^{5} nb\left(x; \frac{1}{2}, 3\right) = \frac{1}{2^3} + \frac{3}{2^4} + \frac{6}{2^5} = \frac{1}{2}.$$

In contrast to this, the binomial probability of three heads in five tosses is

$$\binom{5}{3}\left(\frac{1}{2}\right)^5 = \frac{10}{2^5} = \frac{10}{32}.$$

A careful analysis of the experimental set-up is necessary in order to distinguish between the two situations.

9-3 Poisson Distribution

9-3-1 Genesis as a Limit of the Binomial Distribution: In the binomial distribution, if p is very small, as n becomes increasingly large, it is difficult to evaluate the probabilities. For example, if $X \sim b(x; 100, .001)$, determination of probabilities such as $\Pr(X > 20)$ involves the evaluation of the binomial coefficients

$$\binom{100}{k} \qquad \text{for } k = 21, 22, \ldots, 100$$

and powers of $p = .001$ for corresponding values of k. Such arithmetic is somewhat boggling to the mind. We shall see, however, that as n becomes very large $(n \to \infty)$ with p very small $(p \to 0)$ in such a way that p is proportional to $1/n$, the binomial distribution can be approximated by the *Poisson distribution*.

Consider a binomial r.v. X with parameters p and n. Let $n \to \infty$ and $p \to 0$ such that p is proportional to $1/n$; that is, for each n the probability p depends on n and λ in such a way that $np = \lambda$.

Before proceeding to investigate the limit of the binomial p.f., it may be helpful to review Appendix E-4. With these results in mind, consider $\Pr(X = 0)$ for the binomial distribution

$$\Pr(X = 0) = b(0; n, p) = (1 - p)^n.$$

Taking the limit in the way described above, we have

$$\lim (1 - p)^n = \lim\left(1 - \frac{\lambda}{n}\right)^n,$$

which is $e^{-\lambda}$. A similar consideration of $\Pr(X = 1)$ gives

$$\begin{aligned}\lim \binom{n}{1} p(1 - p)^{n-1} &= \lim n \left(\frac{\lambda}{n}\right)\left(1 - \frac{\lambda}{n}\right)^{n-1}\\ &= \lim \lambda\left(1 - \frac{\lambda}{n}\right)^{n-1}\\ &= \lim \lambda\left(1 - \frac{\lambda}{n}\right)^{n}\left(1 - \frac{\lambda}{n}\right)^{-1}.\end{aligned}$$

The last term on the r.h.s. will tend to 1. Thus, by definition, we have

$$\lim \lambda\left(1 - \frac{\lambda}{n}\right)^n = \lambda e^{-\lambda}.$$

In general, with $p = \lambda/n$,

$$\begin{aligned}\Pr(X = x) &= \frac{n!}{(n - x)!x!} p^x (1 - p)^{n-x}\\ &= \frac{n(n - 1) \cdots (n - x + 1)}{x!} \left(\frac{\lambda}{n}\right)^x\left(1 - \frac{\lambda}{n}\right)^n\left(1 - \frac{\lambda}{n}\right)^{-x}.\end{aligned}$$

For fixed x, as n goes to infinity, the terms $\left(1 - \frac{\lambda}{n}\right)^{-x}$ and $\frac{n(n - 1) \cdots (n - x + 1)}{n^x}$ will both go to 1. Then

$$\lim \Pr(X = x) = \frac{\lambda^x}{x!} e^{-\lambda}.$$

This limiting process suggests a r.v. X with p.f.

$$p(x; \lambda) = \frac{1}{x!} \lambda^x e^{-\lambda} \qquad x = 0, 1, 2, \ldots,$$

which arises as the limit of the binomial distribution. The r.v. X with the p.f. $p(x; \lambda)$ is said to have a Poisson distribution with parameter λ.

The following table gives a comparison of the binomial distribution b(x; 100, .01) and the Poisson distribution p(x; 1). The binomial and Poisson probabilities can be found in Tables 1 and 2, respectively.

x	b(x; 100, .01)	p(x; 1)
0	.3360	.3679
1	.3697	.3679
2	.1849	.1839
3	.0610	.0613
4	.0149	.0153
5	.0029	.0031
6	.0005	.0005
7	.0001	.0001
7+	.0000	.0000

9-3-2 Properties of the Poisson Distribution: Before deriving the mean and variance, it may be useful to show that the individual Poisson probabilities do sum to 1; that is, we wish to evaluate

$$\sum_{x=0}^{\infty} \frac{\lambda^x}{x!} e^{-\lambda}.$$

Recalling that

$$e^{\lambda} = 1 + \lambda + \frac{\lambda^2}{2!} + \frac{\lambda^3}{3!} + \cdots = \sum_{x=0}^{\infty} \frac{\lambda^x}{x!}$$

it can be seen that

$$\sum_{x=0}^{\infty} \frac{\lambda^x}{x!} e^{-\lambda} = e^{-\lambda} \sum_{x=0}^{\infty} \frac{\lambda^x}{x!} = e^{-\lambda} e^{\lambda} = 1.$$

Theorem 9·7: *If $X \sim p(x; \lambda)$, then*

$$E(X) = \lambda \quad \textit{and} \quad Var(X) = \lambda.$$

Proof:

By definition,

$$E(X) = \sum_{x=0}^{\infty} x \frac{\lambda^x}{x!} e^{-\lambda} = e^{-\lambda} \sum_{x=1}^{\infty} x \frac{\lambda^x}{x!} .$$

Substituting $j = x - 1$,

$$E(X) = e^{-\lambda} \sum_{j=0}^{\infty} \frac{(j + 1)\lambda^{j+1}}{(j + 1)!} = e^{-\lambda}\lambda \sum_{j=0}^{\infty} \frac{\lambda^j}{j!}$$

$$= \lambda e^{-\lambda} e^{\lambda} = \lambda.$$

Since the expectation of a binomial random variable is np, this result for the mean of the Poisson distribution is not surprising.

Using an analogous argument, we can find the $E[X(X - 1)]$ to be λ^2. Then recalling that

$$Var(X) = E[X(X - 1)] + E(X) - [E(X)]^2,$$

we have in this case

$$Var(X) = \lambda^2 + \lambda - \lambda^2 = \lambda. \qquad \square$$

Note that for this distribution the mean and variance are equal.

9-3-3 *Using the Poisson Distribution:*

Example 7: Suppose the chance of triplets in human births is .0001. What is the probability of observing at least four sets of triplets in 10,000 human births?

Solution: Here n = 10,000 and p = .0001. We need the probability

$$Pr(X \geq 4) = \sum_{x=4}^{10,000} \binom{10,000}{x} (.0001)^x (1 - .0001)^{10,000-x}.$$

Try to evaluate that expression! Using the Poisson approximation with $\lambda = np = 1$, we have

$$\Pr(X \geq 4) = \sum_{x=4}^{\infty} e^{-1} \frac{(1)^x}{x!} = 1 - \sum_{x=0}^{3} e^{-1} \frac{(1)^x}{x!} = .019.$$

9-4 Hypergeometric Distribution

9-4-1 Origin of the Distribution: Sampling Without Replacement Revisited: Again consider experiments which are made up of a fixed number of trials in which there are two possible outcomes for each trial. To fix ideas, we might think of a box with N balls of which M are red and N - M are blue. We have shown (see Sections 3-5 and 4-3) that if we take a sample of n balls from this box without replacement, then

1. The probability of a red ball is M/N for all draws.
2. The draws are not independent.

Now define a r.v. X = number of red balls in n draws. Is X a binomial random variable? These trials are not Bernoulli since they are not independent; thus, X is not a binomial random variable. Clearly each possible sample of size n contains x red balls and n - x blue balls. From basic counting techniques we know that there are

$$\binom{M}{x}\binom{N - M}{n - x}$$

samples of x red and n - x blue balls. Since there are in all $\binom{N}{n}$ possible samples of size n, then

$$\Pr(X = x) = \frac{\binom{M}{x}\binom{N - M}{n - x}}{\binom{N}{n}}$$

with x = 0, 1, ..., min(M, n). Since these probabilities form the terms of a hypergeometric series, we will call this a hypergeometric distribution with parameters M, N, n and designate it by h(x; M, N, n). The parameters are defined by

M = number of items of interest in population

N = total number of items in population

n = sample size.

9-4-2 Mean and Variance: In this section, we will assume that n is less than M; hence, the value set for X is 0, 1, 2, ..., n.

> Theorem 9·8: *If* $X \sim h(x; M, N, n)$, *then*
>
> $$E(X) = n\frac{M}{N}.$$

Proof (i):

By definition,

$$E(X) = \sum_{x=0}^{n} x \frac{\binom{M}{x}\binom{N-M}{n-x}}{\binom{N}{n}},$$

which can be rewritten as

$$E(X) = \frac{1}{\binom{N}{n}} \sum_{x=1}^{n} \frac{M!}{(x-1)!(M-x)!} \frac{(N-M)!}{(n-x)!(N-M-n+x)!}.$$

Substituting $j = x - 1$ gives

$$E(X) = \frac{M}{\binom{N}{n}} \sum_{j=0}^{n-1} \binom{M-1}{j}\binom{N-M}{n-j-1}.$$

Application of Theorem D·2 yields

$$E(X) = \frac{M}{\binom{N}{n}} \binom{N-1}{n-1} = n\frac{M}{N}.$$

Proof (ii):

As in the binomial case, the expectation here can be found more elegantly by using indicator random variables. Let I_j indicate a success on the jth trial with $j = 1, 2, \ldots, n$. By the equivalence law (Theorem 3·7), $\Pr(I_j = 1) = M/N$ for all j. Clearly

$$X = I_1 + I_2 + \cdots + I_n.$$

Using Theorem 8·3, we have

$$E(X) = E\left(\sum_{j=1}^{n} I_j\right) = \sum_{j=1}^{n} E(I_j) = \sum_{j=1}^{n} \frac{M}{N} = n\,\frac{M}{N}. \qquad \square$$

Theorem 9·9: *If* $X \sim h(x;\, M,\, N,\, n)$, *then*

$$\mathrm{Var}(X) = n\,\frac{M}{N}\left(1 - \frac{M}{N}\right)\left(\frac{N - n}{N - 1}\right).$$

Proof (i):

The proof of this theorem is analogous to proof (i) of Theorem 9·2. The details are left as an exercise.

Proof (ii):

At this point we are well aware of an alternative proof using indicator random variables. We can write

$$X = \sum_{j=1}^{n} I_j.$$

To find the variance of X, we apply Theorem 8·7:

$$\mathrm{Var}(X) = \mathrm{Var}\left(\sum_{j=1}^{n} I_j\right) = \sum_{j=1}^{n} \mathrm{Var}(I_j) + 2 \sum_{i<j}\sum \mathrm{Cov}(I_i, I_j).$$

Since the I_j are indicator random variables,

$$\text{Var}(I_j) = \frac{M}{N}\left(1 - \frac{M}{N}\right) \qquad \text{for all } j.$$

How do we determine $\text{Cov}(I_i, I_j)$? We have proved that

$$\text{Cov}(I_i, I_j) = \Pr(I_i = 1 \text{ and } I_j = 1) - \Pr(I_i = 1)\Pr(I_j = 1),$$

which in this case gives

$$\text{Cov}(I_i, I_j) = \frac{M(M - 1)}{N(N - 1)} - \frac{M}{N}\frac{M}{N}.$$

Thus, combining these results,

$$\begin{aligned}
\text{Var}(X) &= \sum_{j=1}^{n} \frac{M}{N}\left(1 - \frac{M}{N}\right) + 2 \sum_{i<j}\sum \left[\frac{M(M - 1)}{N(N - 1)} - \frac{M}{N}\frac{M}{N}\right] \\
&= n\frac{M}{N}\left(1 - \frac{M}{N}\right) + \frac{2n(n - 1)}{2}\left[\frac{M(M - 1)}{N(N - 1)} - \frac{M^2}{N^2}\right] \\
&= n\frac{M}{N}\left(1 - \frac{M}{N}\right)\left(\frac{N - n}{N - 1}\right).
\end{aligned}$$

□

The following example illustrates how the hypergeometric distribution can be used in acceptance sampling.

Example 8: In large-scale manufacturing it is often impractical to maintain quality control by examining each item. Instead, the product is inspected in lots of k items. Each lot is then accepted or rejected after a sample from it is examined. Suppose the items are produced in lots of 10. Consider the following two procedures of acceptance sampling based on a sample of size two taken without replacement.

I: If both items are nondefective, accept the lot; otherwise, reject it.

II: If both are nondefectives, accept the lot. If both are defective, reject the lot. If, however, one is defective and the other nondefective, a third item is sampled from the lot. The lot is accepted or rejected depending on whether this item is nondefective or defective.

(a) For each procedure determine the probability of accepting the lot as a function of the number of defectives in the lot.

(b) Graph the number of defectives vs. the probability of accepting the lot for each procedure. This graph is called the *operating characteristic curve* of the procedure.

(c) If you were the consumer for this product, which procedure would you prefer? Would your decision be different if you were the manufacturer?

Solution: Let d be the number of defectives in the lot and A the event of accepting the lot. With d = 0, 1, ..., 10, we have

(a) For procedure I

$$\Pr(A_I) = \frac{\binom{10-d}{2}}{\binom{10}{2}}$$

and for procedure II

$$\Pr(A_{II}) = \frac{\binom{10-d}{2}}{\binom{10}{2}} + \frac{\binom{10-d}{1}\binom{d}{1}}{\binom{10}{2}}\,\frac{\binom{9-d}{1}}{\binom{8}{1}}\,.$$

(b)

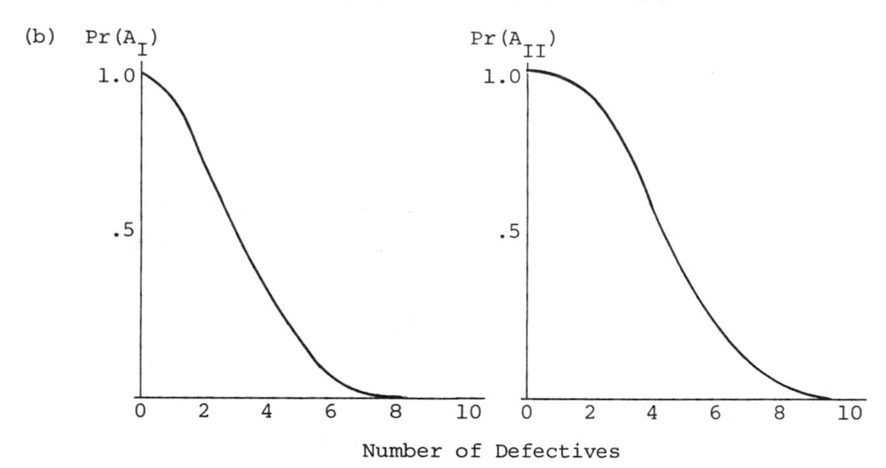

(c) Procedure I accepts less than one-half the time when the lots have as few as three defectives. With procedure II, there is a 50%

chance that lots having as many as five defectives will be accepted. It is quite clear that the decision as to which procedure is better may depend upon whether you are the consumer or manufacturer.

If the lot size had been very large in comparison with the number of items being sampled, how would we determine the probability of accepting the lot? In this case we approximate the hypergeometric distribution by the binomial. This approximation will be investigated in the next section.

9-4-3 A Comparison of the Binomial and Hypergeometric Distributions: The following table summarizes the conditions under which binomial and hypergeometric random variables arise:

Properties of	
Binomial Distribution	*Hypergeometric Distribution*
1. *Series of* n *trials.*	1. *Series of* n *trials.*
2. *Two outcomes* (S *or* F) *for each trial.*	2. *Two outcomes* (S *or* F) *for each trial.*
3. Pr(S) *is the same from trial to trial.*	3. Pr(S) *is the same from trial to trial.*
4. *Trials are independent.*	4. *Trials are not independent.*

An examination of the table shows that the distinguishing feature between the two random variables is whether or not the trials are independent. Thus, if in a hypergeometric type experiment the trials are made "independent," then essentially we would be dealing with a binomial random variable. When will this occur? Intuitively it should seem reasonable that *if the population size is large relative to the sample being drawn*, then the proportion of the two kinds of objects will not change appreciably as the selection proceeds. Effectively, then, sampling without replacement will become equivalent to sampling with replacement. This implies that the conditional probability of success at each draw will approach the unconditional binomial probability of success $p = M/N$. If M and N are large and in a fixed ratio, then

$$\frac{M - a}{N - b} \to p \quad \text{and} \quad \frac{N - M - a}{N - b} \to 1 - p$$

for any constants a and b. Let us now examine the hypergeometric probability function

$$h(x;\ M,\ N,\ n) = \frac{\binom{M}{x}\binom{N - M}{n - x}}{\binom{N}{n}} = \frac{n!}{x!(n - x)!}\left[\frac{M(M - 1)\ \cdots\ (M - x + 1)}{N(N - 1)\ \cdots\ (N - x + 1)}\right]$$

$$\cdot\left[\frac{(N - M)\ \cdots\ (N - M - n + x + 1)}{(N - x)\ \cdots\ (N - n + 1)}\right].$$

Taking the limit, we have

$$\lim h(x;\ M,\ N,\ n) = \frac{n!}{x!(n - x)!}\left(\frac{M}{N}\right)^x\left(1 - \frac{M}{N}\right)^{n-x}$$

$$= \binom{n}{x}p^x(1 - p)^{n-x},$$

which is a binomial probability function with parameters n and p. In summary, then,

$$\lim h(x;\ M,\ N,\ n) = b(x;\ n,\ p).$$

This limit process can also be applied to the mean and variance of the hypergeometric random variable: Let X_h and X_b represent hypergeometric and binomial random variables respectively. Then

$$E(X_h) = n\,\frac{M}{N} = np = E(X_b)$$

and

$$\lim \text{Var}(X_h) = \lim n\,\frac{M}{N}\left(1 - \frac{M}{N}\right)\left(\frac{N - n}{N - 1}\right).$$

Noting that

$$\lim_{N\to\infty} \frac{N - n}{N - 1} = 1,$$

then

$$\lim \text{Var}(X_h) = npq = \text{Var}(X_b).$$

Example 9: A population of N people contains an equal number of men and women. Suppose a sample of size 3 is drawn; what is the probability function of X, the number of women in the sample?

Solution: Recall that

$$h(x; \frac{N}{2}, N, 3) = \frac{\binom{N/2}{x}\binom{N/2}{3-x}}{\binom{N}{3}} \qquad x = 0, 1, 2, 3.$$

Since there are an equal number of males and females, h(0) = h(3) and h(1) = h(2). Let us examine the p.f. for various values of N and see how well it can be approximated by a binomial p.f.

N	Pr(X = 0) = Pr(X = 3)	Pr(X = 1) = Pr(X = 2)
20	.105	.395
50	.117	.383
100	.121	.379
1000	.1246	.3754
∞ (binomial)	.1250	.3750

Clearly, as N increases, the probabilities approach the binomial probabilities.

9-5 Sums of Binomial Random Variables

Returning again to the problem of a marksman shooting at a target, suppose that the probability that Jim hits the target is .3. Let X_1 and X_2 count the number of times he is successful in three trials on Monday and four trials on Tuesday, respectively. Then

$$X_1 \sim b(x_1; 3, .3)$$

$$X_2 \sim b(x_2; 4, .3).$$

Now suppose we are interested in the total number of times Jim hits the target in the seven trials that he performs on these two days. Then T, the total number of successes, can be represented as $T = X_1 + X_2$. Since the value set for X_1 is {0, 1, 2, 3} and that for X_2 is {0, 1, 2, 3, 4}, the value set for T is {0, 1, ..., 7}.

How might we compute the probabiiity that T = 4?

$$\Pr(T = 4) = \Pr\{(X_1 = 0, X_2 = 4) \text{ or } (X_1 = 1, X_2 = 3) \text{ or } (X_1 = 2, X_2 = 2) \text{ or } (X_1 = 3, X_2 = 1)\}. \quad (1)$$

In (1) the event $(X_1 = 4, X_2 = 0)$ has been omitted since it is an impossible event. Assuming that the attempts on the two days are independent,

$$\Pr(X_1 = j, X_2 = k) = \Pr(X_1 = j)\Pr(X_2 = k).$$

Applying this result and noting that the events on the r.h.s. of (1) are m.e., we have

$$\begin{aligned}\Pr(T = 4) &= \Pr(X_1 = 0)\Pr(X_2 = 4) + \Pr(X_1 = 1)\Pr(X_2 = 3) \\ &\quad + \Pr(X_1 = 2)\Pr(X_2 = 2) + \Pr(X_1 = 3)\Pr(X_2 = 1) \\ &= \binom{3}{0}(.3)^0(.7)^3\binom{4}{4}(.3)^4(.7)^0 \\ &\quad + \binom{3}{1}(.3)^1(.7)^2\binom{4}{3}(.3)^3(.7)^1 \\ &\quad + \binom{3}{2}(.3)^2(.7)^1\binom{4}{2}(.3)^2(.7)^2 \\ &\quad + \binom{3}{3}(.3)^3(.7)^0\binom{4}{1}(.3)^1(.7)^3. \end{aligned} \quad (2)$$

The common factor in (2) is $(.3)^4(.7)^3$; hence,

$$\Pr(T = 4) = (.3)^4(.7)^3\left[\binom{3}{0}\binom{4}{4} + \binom{3}{1}\binom{4}{3} + \binom{3}{2}\binom{4}{2} + \binom{3}{3}\binom{4}{1}\right].$$

From Vandermonde's theorem, the term in square brackets is $\binom{7}{4}$. Combining these results gives

$$\Pr(T = 4) = (.3)^4(.7)^3\binom{7}{4}. \quad (3)$$

This expression in (3) suggests that

$$T \sim b(t;\ 7,\ .3).$$

We have considered a particular example. Is this result true in general?

Consider two independent binomial random variables

$$X_1 \sim b(x_1;\ n_1,\ p)$$

$$X_2 \sim b(x_2;\ n_2,\ p).$$

We are interested in the sum of these two random variables

$$T = X_1 + X_2$$

with value set $\{0, 1, \ldots, n_1 + n_2\}$. Note that $T = k$ when $X_1 = i$ and $X_2 = k - i$ for $i = 0, 1, 2, \ldots, k$. Generalizing (1), (2) and (3), we have

$$\begin{aligned}
\Pr(T = k) &= \Pr(X_1 = 0)\Pr(X_2 = k) + \Pr(X_1 = 1)\Pr(X_2 = k - 1) + \cdots \\
&\quad + \Pr(X_1 = k)\Pr(X_2 = 0) \\
&= \binom{n_1}{0} p^0 q^{n_1} \binom{n_2}{k} p^k q^{n_2-k} + \binom{n_1}{1} p^1 q^{n_1-1} \binom{n_2}{k-1} p^{k-1} q^{n_2-k+1} \\
&\quad + \cdots + \binom{n_1}{k} p^k q^{n_1-k} \binom{n_2}{0} p^0 q^{n_2} \\
&= p^k q^{n_1+n_2-k}\left[\binom{n_1}{0}\binom{n_2}{k} + \binom{n_1}{1}\binom{n_2}{k-1} + \cdots + \binom{n_1}{k}\binom{n_2}{0}\right].
\end{aligned}$$

The term in the square brackets is

$$\sum_{i=0}^{k} \binom{n_1}{i}\binom{n_2}{k-i} = \binom{n_1 + n_2}{k};$$

hence,

$$\Pr(T = k) = p^k q^{n_1+n_2-k}\binom{n_1 + n_2}{k}$$

or

$$T \sim b(t;\ n_1 + n_2,\ p).$$

This result depends upon p, the probability of success, being the same for both binomial random variables. Clearly if p is not the same, the expression $p^k q^{n_1+n_2-k}$ will not be a common factor of each term; T will <u>not</u> be a binomial random variable.

Using mathematical induction this result can easily be extended to r <u>independent</u> series of trials each based on the same probability of success. Let

$$X_i \sim b(x_i;\ n_i,\ p)$$

for i = 1, 2, ..., r and with X_j and X_k independent for all j and k. then if $T = \sum_{i=1}^{n} X_i$,

$$T \sim b(t;\ \sum_{i=1}^{r} n_i,\ p).$$

Proof of this result is left as an exercise.

We have seen in Section 9-3 that for large n and small p, the binomial distribution can be approximated by the Poisson with parameter np. In the archery example suppose that Jim is a novice with p = .01. In order to improve his performance, he practices 100 times on Monday and 200 times on Tuesday. Now

$$X_1 \sim b(x_1;\ 100,\ .01) \overset{\text{approx.}}{\sim} p(x_1;\ 1)$$

$$X_2 \sim b(x_2;\ 200,\ .01) \overset{\text{approx.}}{\sim} p(x_2;\ 2).$$

We have seen that

$$T = X_1 + X_2 \sim b(t;\ 300,\ .01) \overset{\text{approx.}}{\sim} p(t;\ 3).$$

This seems to suggest the general result: If Y_1 and Y_2 are independent Poisson random variables with parameters λ_1 and λ_2, then

$$T = Y_1 + Y_2 \sim p(t; \lambda_1 + \lambda_2).$$

As before, consider

$$\Pr(T = k) = \Pr(Y_1 = 0)\Pr(Y_2 = k) + \Pr(Y_1 = 1)\Pr(Y_2 = k - 1) + \cdots + \Pr(Y_1 = k)\Pr(Y_2 = 0)$$

$$= \frac{e^{-\lambda_1}\lambda_1^0}{0!}\frac{e^{-\lambda_2}\lambda_2^k}{k!} + \frac{e^{-\lambda_1}\lambda_1^1}{1!}\frac{e^{-\lambda_2}\lambda_2^{k-1}}{(k-1)!} + \cdots + \frac{e^{-\lambda_1}\lambda_1^k}{k!}\frac{e^{-\lambda_2}\lambda_2^0}{0!}.$$

Taking out the common factor $e^{-(\lambda_1+\lambda_2)}$, we have

$$\Pr(T = k) = e^{-(\lambda_1+\lambda_2)}\left[\frac{\lambda_1^0\lambda_2^k}{0!k!} + \frac{\lambda_1^1\lambda_2^{k-1}}{1!(k-1)!} + \cdots + \frac{\lambda_1^k\lambda_2^0}{k!0!}\right].$$

Multiplying the r.h.s. by k!/k! gives

$$\Pr(T = k) = \frac{e^{-(\lambda_1+\lambda_2)}}{k!}\left[\binom{k}{0}\lambda_1^0\lambda_2^k + \binom{k}{1}\lambda_1^1\lambda_2^{k-1} + \cdots + \binom{k}{k}\lambda_1^k\lambda_2^0\right].$$

The term in the square brackets is the expansion of $(\lambda_1 + \lambda_2)^k$, hence,

$$\Pr(T = k) = \frac{e^{-(\lambda_1+\lambda_2)}(\lambda_1 + \lambda_2)^k}{k!} \qquad k = 0, 1, 2, \ldots$$

or

$$T \sim p(t; \lambda_1 + \lambda_2).$$

A generalization of this to the sum of r independent Poisson random variables is left as an exercise.

If Jim improves with practice so that his probability of success on the second day is .02, then

$$X_1 \sim b(x_1;\ 100,\ .01)$$

$$X_2 \sim b(x_2;\ 200,\ .02).$$

We have seen that

$$T = X_1 + X_2$$

is not binomial. Thus it would appear that T cannot be approximated by the Poisson. Since

$$X_1 \overset{\text{approx.}}{\sim} p(x_1;\ 1)$$

$$X_2 \overset{\text{approx.}}{\sim} p(x_2;\ 4),$$

we can use the above result and write

$$T \overset{\text{approx.}}{\sim} p(t;\ 5).$$

Thus if

$$X_i \sim b(x_i;\ n_i,\ p_i)$$

with each n_i large and p_i small, we have

$$T = \Sigma X_i \overset{\text{approx.}}{\sim} p(t;\ \Sigma p_i n_i).$$

Is it necessary that we be able to approximate each binomial component by a Poisson? That is, what happens if each n_i is small? Suppose Jim practiced 20 times each day for five days with $p = .01$. Clearly

$$X_i \sim b(x_i;\ 20,\ .01)$$

for $i = 1, 2, \ldots, 5$. Although each X_i cannot be approximated by a Poisson,

$$T = \Sigma X_i \sim b(t;\ 100,\ .01) \overset{\text{approx.}}{\sim} p(t;\ 1).$$

In general, if

$$X_i \sim b(x_i; n_i, p)$$

with Σn_i large, then

$$T = \Sigma X_i \overset{\text{approx.}}{\sim} p(t; p \Sigma n_i).$$

Problems

1. Let $X \sim b(x; n, p)$.
   (a) For $n = 6$, $p = .2$, find (i) $Pr(X > 3)$, (ii) $Pr(X \geq 3)$, (iii) $Pr(X < 2)$.
   (b) For $n = 15$, $p = .8$, find (i) $Pr(X \leq 2)$, (ii) $Pr(X \geq 12)$, (iii) $Pr(X = 8)$.
   (c) For $n = 10$, find p so that $Pr(X \geq 8) = .6778$.

2. Let X be a binomial random variable with $\mu = 6$ and $\sigma^2 = 2.4$. Find
   (a) $Pr(X > 2)$.
   (b) $Pr(2 < X < 8)$.
   (c) $Pr(X \leq 8)$.

3. How many Bernoulli trials with probability of success equal to .01 must be performed in order that probability of at least one success is 1/2 or more?

4. Consider a game in which a person wins if a 1 or 2 appears on a balanced die and loses if a 3, 4, 5 or 6 appears. In three rolls, find the probability of
   (a) At least two wins.
   (b) At most two wins.
   (c) Exactly two wins.

5. As a part of a government project, a research worker is studying the annual salary of families. In the population the proportion of families having an annual salary greater than \$15,000 is p.

5. Suppose a sample of n families is taken with replacement and the research worker notes for each family whether the annual salary is >\$15,000 or ≤\$15,000. Show that the quantity which is being recorded is an indicator random variable. What is its probability function?

6. Continuing Question 5, suppose the researcher is interested in S, the total number of families in the sample whose income is >\$15,000.
   (a) Find E(S) and Var(S).
   (b) What is the distribution of S?
   (c) For a sample of size 15 and p = .40, find Pr(X ≥ 12).

7. Extending Quesions 5 and 6, consider the proportion P of the families in the sample whose annual salary is >\$15,000.
   (a) How can P be defined in terms of S?
   (b) Find E(P) and Var(P).
   (c) For n = 15 and p = .40, find the probability that P ≥ 4/5. Compare this with your answer to Question 6(c).
   (d) Consider the results of part (b). What in general can you say about the variance of P when n becomes large? What is the practical implication of this result?

8. Find the expectation and variance of Y, the number of failures, in n binomial trials with p as the probability of success.

9. Suppose in a series of five binomial trials you observe three successes. Consider the binomial probabilities

$$b(3;\ 5,\ p) = \binom{5}{3} p^3 (1 - p)^2$$

as a function of p. Using Table 1, show that this function is maximized when p = 3/5. Can you suggest a generalization of this result for b(x; n, p)?

10. A standard treatment for a certain disease leads to cures in one-fourth of the cases. A new treatment is devised which is said to produce cures in three-fourths of the cases. The new

10. treatment is tried out on 10 persons having the disease with three possible courses of action to be taken:

    (i) If seven or more are cured, the new treatment is adopted.

    (ii) If three or fewer are cured, the new treatment will be discarded.

    (iii) If four, five or six are cured, further investigation will be undertaken.

    (a) Find the probability of each of these courses of action when the new treatment is no more effective than the standard; that is, $p = 1/4$.

    (b) Find the probability of each of these courses of action if the new treatment is actually as effective as claimed.

    (c) Comment on the implications of the results obtained in (a) and (b).

11. (a) Show that

$$b(k; n + 1, p) = pb(k - 1; n, p) + (1 - p)b(k; n, p).$$

    (b) Use the identity in part (a) to show how Table 1 can be extended to $n = 16$.

12. In comparing two drugs M and N, a doctor decides to give some patients drug M and some patients drug N. Prior to seeing each patient, he tosses a balanced coin. If a head appears, he gives drug M; otherwise he gives drug N. If the experiment involves 15 persons,

    (a) What is the probability that all patients will get the same drug?

    (b) What is the probability that the assignment of the drugs will alternate?

13. For the purpose of a certain experiment, mice are classified as uniformly colored (U) or piebald (P). A geneticist studies 100 litters with eight mice in each and counts the number of piebald mice in each litter.

Number of (P)	0	1	2	3	4	5	6	7	8
Observed Frequency	0	0	5	14	21	29	20	9	2

Genetic theory predicts that the number of piebald in a litter of eight will be distributed as a binomial random variable with $n = 8$ and p unknown.

(a) Estimate p on the basis of the above data.

(b) Calculate the expected frequencies on the basis of your estimate. Does the binomial model seem appropriate?

(c) Previous studies have suggested the value $p = .7$. Recalculate the expected frequencies corresponding to $p = .7$. Does this value seem reasonable in light of the above data?

14. Prove the following recursive formulas:

(a) $b(k + 1; n, p) = \frac{n - k}{k + 1} \frac{p}{1 - p} b(k; n, p)$

(b) $p(k + 1; \lambda) = \frac{\lambda}{k + 1} p(k; \lambda)$

How might these results be used for constructing tables of the appropriate probability functions?

15. In order to estimate the total number of fish in a lake, a capture-recapture experiment was conducted. One hundred fish were captured, tagged and then returned to the lake. In a future sample of 25 fish, five were found to be tagged.

(a) Estimate the total number of fish in the lake. What assumptions are you making?

(b) If in fact there were 1000 fish in the lake, what is the chance of observing five or fewer tagged fish in the sample of 25?

16. The probability that a person dies from a particular type of heart disorder is .001.

(a) Find the probability that fewer than five of 2000 people with the disorder will die.

16. (b) Determine the interval $\mu \pm 2\sigma$ and place an interpretation on this interval using Chebyshev's inequality.

17. A survey is being conducted to estimate the proportion of car owners who have two or more cars. How large should the sample be to ensure that, with probability .9, the estimate is within .05 of the true proportion?

18. The probability that a boy will hit the target is 1/4. How many shots should he take to ensure that, with a probability of at least .5, the proportion of hits will not differ from the true proportion by more than 1/4?

19. A bird cage has three healthy plump canaries and two scrawny ones. After a struggle Felix, the cat, nabs the three healthy birds and prepares to gobble them. Just at that moment his master appears and says, "Felix, repeat that feat twice in three tries and you'll have the birds for dinner." What is the probability that Felix has the birds for dinner?

20. To test the effects of a drug on the response rate, the following experiment was conducted: Ten rats were trained to respond to a stimulus by rewarding them. Then each rat was given the drug and the stimulus was applied. The response rate for each rat was measured. After a suitable period of time to allow the drug effect to wear off, the stimulus was again applied and the reaction time measured. The results were recorded as

S--Rat's response was slower when drug was applied.

F--Rat's response was faster when drug was applied.

(a) If the drug has no effect, what is the chance of recording an S for a single rat?

(b) If the drug has no effect, what is the distribution of X, the number of S's when ten rats are observed? (State all necessary assumptions.)

(c) If you observe $X = 9$, what would you conclude?

21. The proportion of families subscribing to newspaper A in a certain city is believed to be 60%. Suppose a random sample of 10 families shows that three or fewer families subscribe to paper A.
    (a) Newspaper B argues that paper A's claim is not viable. Based on this sample what can you conclude?
    (b) Find the probability that $X \leq 3$ for $p = .3$, $.4$ and $.5$. How might these probabilities be used in settling the dispute between the two papers?

22. The administration of a state university has set up a committee to investigate the claim that 70% of the students oppose a fee increase. Fifteen students are to be interviewed. The claim will be accepted if X, the number of students opposing, lies between 8 and 12.
    (a) What is the probability of rejecting the claim when in fact it is true?
    (b) What is the probability that the claim will be accepted when in fact only 50% of the students oppose it?
    (c) Discuss these results.

23. In a small experiment, six children of similar age and background were chosen and divided into two groups of three. Each group was taught to read using one of two different methods (I and II). After a six-month period, each child was tested giving the results:

Child Number	Method I	Method II
1	65 (5)	69 (4)
2	54 (6)	85 (1)
3	71 (3)	79 (2)

(The number in parentheses gives the rank of that child.) Suppose we start with the hypothesis that the two methods are equally effective and hence (it seems reasonable to assume that) the ranks of the children taught with method II are a

23. random selection from the six ranks available. Let T be the sum of the ranks of the method II group.

    (a) Find the probability function of T assuming the hypothesis discussed above is true.

    (b) Find E(T) and Var(T).

    (c) What values of T would make you think that, in fact, method II is more effective? Does the observed value (T = 7) seem unusual with regards to the hypothesis?

24. In studying the behavior of changes in the stock market, an economist calculates an index at the end of each day. Experience shows that each day the index moves either up or down. He studies the index in four-day periods. He notices that in more than half the cases there are at least two changes in direction. In fact, studying 20 graphs, he finds that in 13 instances there were at least two changes in direction. To explain this phenomenon he theorizes that there is an economic force acting which, once the index starts advancing, causes it to reverse and vice versa. As a statistician, you suggest alternatively that the index is moving up and down at random within each recording period.

    (a) Make a list of the possible outcomes for a single four-day period.

    (b) If, as hypothesized, these are all equally likely, what is the chance that in a given case there will be two or more changes in direction?

    (c) If we assume that each case is independent, why can the number of cases X with two or more changes be considered as a binomial random variable?

    (d) Find $P(X \geq 13)$ on the basis of the assumption of randomness. Do the observed data seem to refute the hypothesis of randomness?

25. Canned fruit juices come in cases of 12 cans and are shipped in lots of eight cases. To examine the quality of a shipment, a company uses the following procedure:

    (1) Choose a case at random from the eight and then

    (2) Choose two cans at random from the chosen case.

    If either of the chosen cans does not meet rigid standards, then the whole shipment is reexamined using a more elaborate procedure. If both the cans are acceptable, then the whole lot is shipped. Suppose of the eight cases, six are perfect, one has one damaged can, and the other has 12 damaged cans.

    (a) What is the chance the lot is declared acceptable?

    (b) Suppose as an alternative procedure, we choose two cans at random from the whole lot and reject the lot if either is defective. What is the chance the lot is accepted in this case?

    (c) Which scheme do you prefer in this case? Is there any siutation in which you would change your opinion?

26. A woman who is cooking two pounds of peas selects 10 peas at random in order to decide whether the peas are sufficiently cooked. If none of the 10 peas is insufficiently cooked, she will decide that all the peas in the pan are fully cooked.

    (a) Draw the operating characteristic curve for her sampling scheme.

    (b) If 10% of the peas are insufficiently cooked, calculate the probability that she will say that the peas in the pan are completely cooked.

    (c) Comment critically on her sampling scheme. How might it be altered?

27. A parachute factory manufactures in lots of 500. Samples of size 5 are tested: If all are satisfactory, the lot is passed. Let p be the proportion of defective parachutes in the lot.

    (a) What assumptions must be made in order that T, the number of defectives in the sample, has a binomial distribution?

27. (b) If T is binomial, write down Pr(T = 0) as a function of p and calculate Pr(T = 0) for p = .1, .2, .3, .4, .5.

    (c) Graph Pr(T = 0) vs. p.

    (d) If you were a parachute buyer, would you regard this as a satisfactory testing procedure? Why or why not?

    (e) If you sampled 150 of the 500 in the lot, would you be willing to assume that T is a binomial variable? In this case, what is the distribution of T?

28. Suppose that in flight airplane engines fail with a probability q independently from engine to engine. A plane makes a successful journey if at least 1/2 of its engines run. For what values of q is a two-engine plane preferable to a four-engine plane? Demonstrate your results both algebraically and graphically.

29. Show that

    (a) $b(k; n, p) = b(n - k; n, 1 - p)$.

    (b) $\sum_{k=r}^{n} b(k; n, p) = 1 - \sum_{k=n-r+1}^{n} b(k; n, 1 - p)$.

    When might these results be useful?

30. In a game of chance John tosses a balanced coin a fixed number of times. He receives a prize if he gets exactly five heads. Before tossing the coin, he is allowed to choose the number of trials. What number should he choose in order to maximize his probability of winning?

31. Let X have the p.f.

$$p(x) = c\left(\frac{1}{4}\right)^x \qquad x = 0, 1, 2, \ldots$$

    (a) Find c so that this is a proper probability function.

    (b) Determine E(X) and Var(X).

32. Suppose Z is distributed as

$$f(z) = kr^z \qquad z = 0, 1, 2, \ldots$$

with k and r as constants such that $0 < r < 1$.

32. (a) Determine k so that this is a proper p.f.
    (b) Find $E(Z)$ and $Var(Z)$.

33. A large lot of transistors has 10% defectives. Let N be the number of draws necessary to find the first defective.
    (a) Stating all the necessary assumptions, find $Pr(X = k)$ for $k = 1, 2, \ldots$
    (b) Find the probability that more than t draws are needed; that is, $Pr(X > t)$.
    (c) Find $Pr(X = j \mid X > t)$ for $j = t + 1, t + 2, \ldots$
    (d) Compare the results of (a) and (c).

34. Let X be the number of loaves of bread sold on a given day. If the p.f. of X is

$$f(x) = \left(\frac{1}{2}\right)^{x+1} \qquad x = 0, 1, 2, \ldots$$

    (a) Show that f(x) is a proper p.f.
    (b) Find the probability that an even number of loaves are sold.
    (c) Find the probability that more than 10 loaves are sold.

35. John and Bill are playing a series of ping pong games. The probability that Bill wins on any game is 3/4. Bill vows that he will play until he wins twice.
    (a) Let X be the number of games needed until Bill wins twice. Find the p.f. of X, stating clearly all assumptions made.
    (b) Show that the p.f. of X is a proper p.f.
    (c) Find $E(X)$ and $Var(X)$.

36. In an oral examination a contestant answers true-false type questions. He continues to answer until he gets eight correct answers. Assuming he guesses the answer on each question,
    (a) What is the probability that
        (i) He needs 20 questions?
        (ii) He needs more than 20 questions?
    (b) What is the expected number of questions he would need?

37. You and two of your friends toss coins to decide who will pay for coffee. Each of you tosses a coin and the person whose coin does not match pays for the coffee. If all three coins turn up the same, they are tossed again.
    (a) Find the probability that fewer than four repetitions are needed to decide who pays for the coffee.
    (b) How many repetitions would you expect would be needed to decide who pays for the coffee?

38. A research scientist is inoculating rabbits, one at a time, with a disease until he finds two rabbits which develop the disease.
    (a) If the probability of contracting the disease is 1/2, what is the probability that eight rabbits are needed?
    (b) Now suppose the rabbits have been immunized against the disease so that the probability of their developing the disease is only 1/6. What is the probability that eight rabbits are needed?

39. There has been much concern recently about abandoned cars on highways. Suppose that the number of cars abandoned in a week on a particular highway has a Poisson distribution with parameter $\lambda = 2$.
    (a) It costs the provincial government $100 per car to tow away and dispose of an abandoned car. What is the expected cost per week to the provincial government to dispose of cars abandoned on the given highway?
    (b) What is the most probable amount of money that the provincial government must pay to dispose of cars abandoned on the highway in a given week?
    (c) How probable is it that the provincial government will have to spend more than $400 in a given week to dispose of cars abandoned on this highway?

40. Assume that the number of baseball gloves purchased in a store during a week's time follows a Poisson distribution with $\lambda = 2$. How many gloves should the merchant stock so that he will be able to supply the demand with probability .95?

41. Using the results of Section 9-5, prove by mathematical induction that

    (a) If X_i, for $i = 1, 2, \ldots, k$, are k independent binomial random variables with probability function $b(x_i; n_i, p)$, then $T = \Sigma X_i$ is distributed as $b(t; \Sigma n_i, p)$.

    (b) If Y_i, for $i = 1, 2, \ldots, r$, are r independent Poisson random variables with probability function $p(y_i; \lambda_i)$, then $S = \Sigma Y_i$ is a Poisson random variable with parameter $\Sigma \lambda_i$.

42. Suppose a typist makes approximately one error in every 50 words typed.

    (a) If we examine 10 examples of 10 words each, find the probability of one or fewer errors. Be sure to state clearly any assumptions and/or approximations which you are using.

    (b) Now suppose we examine one typed passage containing 100 words. Find the probability of one or fewer errors. Is the result the same as in (a)? Why or why not?

43. A senior high school has four classes with 350, 300, 250 and 200 students, respectively. It is felt that approximately 10% of each class are opposed to a new type of grading scheme. If samples of size 35, 30, 20 and 15, respectively, are taken (WOR) from each class, find the probability of 10 or more persons opposing the grading scheme. What assumptions and/or approximations have you used?

44. Machines A and B produce defectives at the rate of 1% and 2%, respectively. If a sample of 100 is taken from each machine, find the probability that the total number of defectives will be more than one. What assumptions and/or approximations have been used?

45. Find the expectation and variance of T and of S as defined in Problem 41.

46. In each of the following experimental situations,

   (i) Identify the type of random variable being described.

   (ii) Give the probability function.

   (iii) State carefully all assumptions made.

   (iv) Indicate appropriate approximations where possible.

   (a) Three percent of a community has a particular blood type. We select five donors (WR) and note the number of donors who have this particular blood type.

   (b) A shipment of 5000 new television sets has 400 with some minor defects and the remaining are in perfect condition. A consumer testing agency randomly chooses 25 sets for inspection and observes the number of defectives.

   (c) Fish are caught one at a time and for each fish caught, the species is noted and the fish is then returned to the pond. This process is continued until 10 fish of the species of interest have been caught. It is known that 30% of the fish in the pond are of this particular species. We are interested in the total number of fish caught.

   (d) In a day's production of 1000 items, 400 were produced by the first shift and 600 by the second shift. A random sample of five items is selected and we note the number in the sample produced on the first shift.

   (e) The proportion of customers who have colored (as opposed to black) telephones is .001. In a random sample of 100 from a population of 100,000, count the number of customers with colored telephones.

Appendix A

Summation and Subscripts

Frequently we may be interested in finding the sum of several observations. For example, in a class of 50 students, we may wish to determine the average student age. To do this we would record the age of each student, sum the ages and divide by 50. How can we express this series of operations in a general and compact way? Let x_1 be the age of the first student on the class roster, x_2 the age of the second, and so on with x_{50} representing the age of the fiftieth student. The sum then is

$$x_1 + x_2 + \cdots + x_{50},$$

where $\cdots$ indicates "and so on." Using the Greek letter Σ (sigma) as a symbol for summation, we can obtain an even more compact representation:

$$\sum_{i=1}^{50} x_i.$$

This expression is read "sum of x sub i where i goes from 1 to 50." The letter i under the summation sign Σ is called the index of summation. It takes on all integral values from its lower limit written below the sigma to its upper limit written above the sigma. Any letter may be used for the index, e.g., i, j, k or ℓ.

Using this notation, we can then find the average as

$$\text{Average} = \frac{1}{50} \sum_{i=1}^{50} x_i.$$

To make this expression more general, suppose the class consisted of n students, then the average would be

$$\text{Average} = \frac{1}{n} \sum_{i=1}^{n} x_i .$$

Sometimes we will find it convenient to omit the limits of summation and simply write Σx_i. This means that the summation extends over all values of i under discussion.

In dealing with summations the following rules may be useful.

Rule 1: *The summation of an expression which is the sum of two or more terms is the sum of their separate summations:*

$$\sum_{i=m}^{n} (x_i + y_i) = \sum_{i=m}^{n} x_i + \sum_{i=m}^{n} y_i .$$

Proof:

The left-hand side (l.h.s.) can be written as

$$x_m + y_m + x_{m+1} + y_{m+1} + \cdots + x_n + y_n,$$

which can be regrouped to give

$$x_m + x_{m+1} + \cdots + x_n + y_m + y_{m+1} + \cdots + y_n .$$

Using the summation symbol we have

$$\sum_{i=m}^{n} x_i + \sum_{i=m}^{n} y_i . \qquad \square$$

Rule 2: *The summation of a constant (a quantity not involving the summation index) multiplied by a variable is the same as the constant multiplied by the summation:*

$$\sum_{i=m}^{n} cx_i = c \sum_{i=m}^{n} x_i.$$

Proof:

The l.h.s. here can be written as

$$cx_m + cx_{m+1} + \cdots + cx_n.$$

Since c is a common factor, we can write

$$c(x_m + x_{m+1} + \cdots + x_n)$$

or in summation notation

$$c \sum_{i=m}^{n} x_i. \quad \square$$

Rule 3: *The summation of a constant is equal to the product of the constant and the number of integers from the lower limit to the upper limit of the summation:*

$$\sum_{i=m}^{n} c = (n - m + 1)c.$$

Proof:

If each x_i in $\sum_{i=m}^{n} x_i$ is equal to c, we have

$$x_m + x_{m+1} + \cdots + x_n = c + c + \cdots + c.$$

There would be n - m + 1 c's on the r.h.s.; hence, we have

$$\sum_{i=m}^{n} c = (n - m + 1)c. \qquad \square$$

Examples of Summation Notation

Example A-1: If $x_1 = -2$, $x_2 = 7$, $x_3 = 5$, find (a) $\sum_{i=1}^{3} x_i$, (b) $\sum_{k=1}^{3} x_k^2$, (c) $\left[\sum_{j=1}^{3} x_j\right]^2$, (d) $\sum_{h=1}^{3} 3x_h$, (e) $\sum_{j=1}^{2} (x_j + x_{j+1})$, (f) $\sum_{j=1}^{3} x_j - 3$, (g) $\sum_{k=1}^{3} (x_k - 3)$.

Solutions:

(a) $\sum_{i=1}^{3} x_i = -2 + 7 + 5 = 10$

(b) $\sum_{k=1}^{3} x_k^2 = (-2)^2 + (7)^2 + (5)^2 = 78$

(c) $\left[\sum_{j=1}^{3} x_j\right]^2 = [-2 + 7 + 5]^2 = 100$

(d) $\sum_{h=1}^{3} 3x_h = 3 \sum_{h=1}^{3} x_h = 3(10) = 30$

(e) $\sum_{j=1}^{2} (x_j + x_{j+1}) = (x_1 + x_2) + (x_2 + x_3) = -2 + 7 + 7 + 5 = 17$

(f) $\sum_{j=1}^{3} x_j - 3 = 10 - 3 = 7$

(g) $\sum_{k=1}^{3} (x_k - 3) = \sum_{k=1}^{3} x_k - \sum_{k=1}^{3} 3 = 10 - 3(3) = 1$

or $\sum_{k=1}^{3} (x_k - 3) = (-2 - 3) + (7 - 3) + (5 - 3) = 1$

Example A-2: Express the following sum as a simple function of n

$$\sum_{k=1}^{n} [(k + 1)^2 - k^2].$$

Solution: Writing this expression in its expanded form, we get

$$\sum_{k=1}^{n} [(k + 1)^2 - k^2] = [2^2 - 1^2] + [3^2 - 2^2] + \cdots + [n^2 - (n - 1)^2] + [(n + 1)^2 - n^2].$$

The r.h.s. of this expression can be written as

$$\{2^2 + 3^2 + 4^2 + \cdots + n^2 + (n + 1)^2\} - \{1^2 + 2^2 + 3^2 + \cdots + (n - 1)^2 + n^2\},$$

which after simplification becomes

$$(n + 1)^2 - 1^2 = n(n + 2).$$

Therefore,

$$\sum_{k=1}^{n} [(k + 1)^2 - k^2] = n(n + 2).$$

Example A-3: Express the sum of the first n natural numbers as a function of n; i.e., evaluate $\sum_{k=1}^{n} k$ as a function of n.

Solution: From Example A-2, we have $\sum_{k=1}^{n} [(k + 1)^2 - k^2] = n(n + 2)$.

The l.h.s. of this equation can also be written as

$$\sum_{k=1}^{n} [k^2 + 2k + 1 - k^2] = \sum_{k=1}^{n} (2k + 1) = 2 \sum_{k=1}^{n} k + n.$$

From Example A-2 this is also equal to n(n + 2); hence,

$$2 \sum_{k=1}^{n} k + n = n(n + 2).$$

Solving for $\sum_{k=1}^{n} k$, we get

$$\sum_{k=1}^{n} k = \frac{\{n(n + 2) - n\}}{2} = \frac{n}{2} \{n + 2 - 1\} = \frac{n(n + 1)}{2} .$$

Appendix B

Set Theory

Since the concepts of set theory are basic to an understanding of elementary probability and statistics, this section of the appendices will provide a brief review of set notation and the algebra of sets.

What do you think of when the word set is mentioned? Perhaps you think of a group, aggregate, class or collection. Since we will be using the concept of a set as a fundamental notion, we will find the following formal definition useful:

> Definition B·1: *A set is any well-defined collection of distinct objects.*

This definition makes it clear that a set is identified by its members. We shall refer to the members of the set as elements. If A represents a set and x is some element in A, then we write

$$x \in A$$

and read this as "x is an element of A." Obviously, if y is an element which does not belong to A, then we write

$$y \notin A$$

and read this as "y is not an element of A."

We will find that there are two quite useful ways of specifying a set: listing its members or giving a rule which characterizes the elements. For example, consider the list A = {1, 2, 3, 5}. This denotes a set A, which consists of the elements 1, 2, 3 and 5. Frequently, the rule method may be more concise:

$$B = \{x \mid x \text{ is a prime number less than seven}\}.$$

Again the curly brackets designate a set and the vertical bar is read "such that."

Both set A and set B have at least one member. Definition B·1, however, does not specify that a set must have members. Suppose that G is the set of all girls attending a boys' school; then the set G has no members.

Definition B·2: *The null set is a set which has no members.*

The null set is symbolized as $\emptyset$ or { }.

Have you realized that set A and set B are the same? When can we say that two sets are the same?

Definition B·3: *Two sets A and B are equal if and only if the sets contain exactly the same elements.*

If two sets are equal, we can write A = B.

Now let the set C be defined as

$$C = \{1, 2, 5, 7, 11\}.$$

Every element of A is an element of C. We say that A is a subset of C or A is contained in C.

Definition B·4: *A set A is called a subset of C if every element in A is an element of C. If C contains at least one element that is not an element of A, then A is called a proper subset of C.*

Symbolically, subsets are denoted by $A \subset C$; that is, A is included in C.

As a consequence of Definition B·4, we see that

1. Two sets A and B are equal if and only if $A \subset B$ and $B \subset A$.
2. Since the null set has no elements, it is a subset of every set: $\emptyset \subset A$ for all A.
3. Every set A is a subset of itself: $A \subset A$.

Suppose we consider a set $F = \{e, f, g\}$. How many subsets can be generated from F?

$$\{\ \}, \{e\}, \{f\}, \{g\}, \{e, f\}, \{e, g\}, \{f, g\}, \{e, f, g\}.$$

From the set F, which has three distinct elements, 2^3 or 8 subsets can be generated. Can you show that 2^k subsets can be generated from a set with k distinct elements?

In discussing sets it may be useful to consider an underlying reference set. We shall assume that all sets under discussion are subsets of some fixed *universal set* symbolized by Ω. The meaning of this universal set may change as our frame of reference changes. For example, in some cases the universal set may be the set of natural numbers or the set of real numbers or even the set of all rooms in your house. In Chapter 2 it becomes clear that, in probability theory, the sample space S is the reference set.

In arithmetic we know that numbers are combined by using such operations as addition and multiplication. What operations do we use for combining sets?

Definition B·5: *Let* A *and* B *be two subsets of* Ω. *Then the set of elements which belongs to* A *or to* B *or both to* A *and* B *is the* union of A and B *and is symbolized by* $A \cup B$.

Definition B·6: *Let* A *and* B *each be a subset of* Ω. *Then the* intersection of A and B *is the set of all emements which belong both to* A *and to* B. *The intersection is denoted by* $A \cap B$.

Sometimes it is helpful to represent sets and set operations by means of Venn diagrams. In Venn diagrams the universal set is represented by a rectangle and the subsets of Ω are depicted as various regions of the rectangle. Note that Venn diagrams are not proofs. They only help in visualizing the relationship between sets. For example, the *union* of A and B and of the *intersection* of A and B can be depicted by means of Venn diagrams:

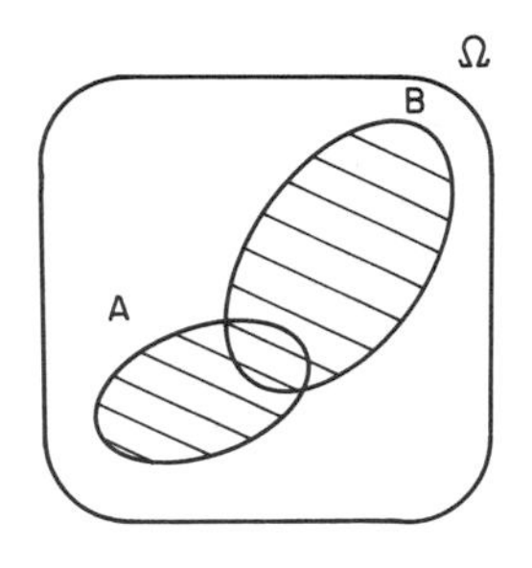

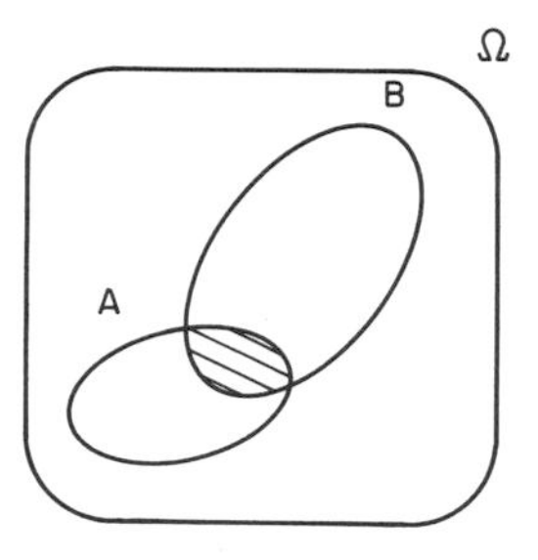

$A \cup B = \{x \mid x \in A \text{ or } x \in B\}$ $\quad$ $A \cap B = \{x \mid x \in A \text{ and } x \in B\}$

If A and B have no elements in common, we say that A and B are *disjoint*; that is, $A \cap B = \emptyset$.

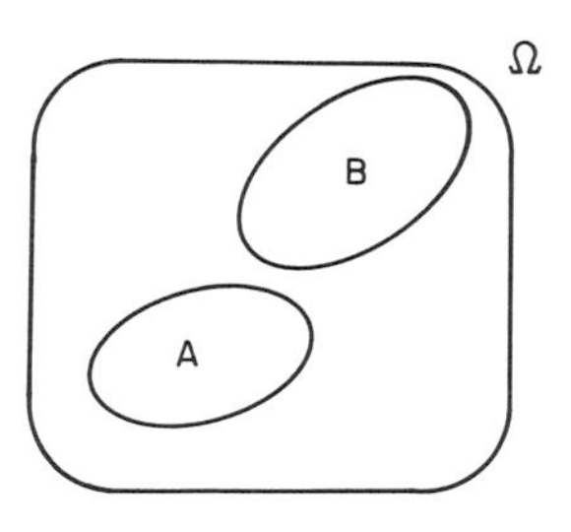

$A \cap B = \emptyset.$

Another set in which we may have particular interest is the set of all elements in Ω which do not belong to A. We call this the complement of A and represent it by $\bar{A}$.

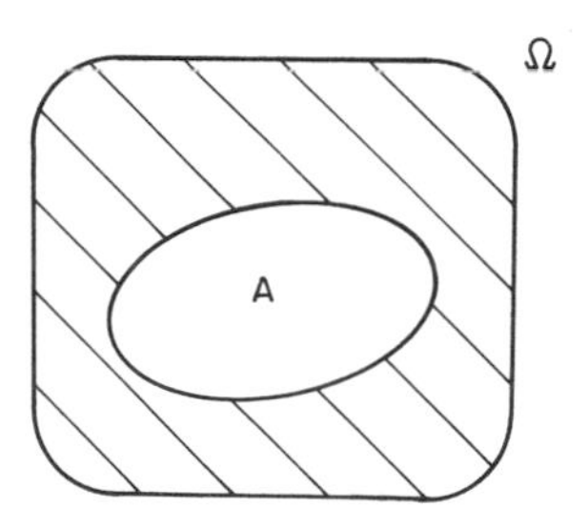

$$\bar{A} = \{x \mid x \notin A \text{ and } x \in \Omega\}$$

Suppose that instead of considering the complement of A with respect to Ω we are interested in the relative complement of A with respect to B. This is often called the difference of B and A; that is, B - A.

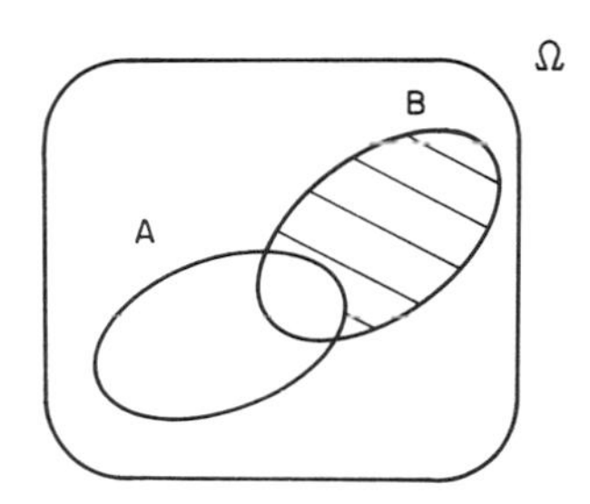

$$B - A = \{x \mid x \in B \text{ and } x \notin A\}$$

What general statements can be made about the interrelationships between intersections, complements and unions of sets? The following list is a review of the *algebra of sets*, which we shall use without proof:

Identity laws:	$A \cup \emptyset = A$	$A \cup \Omega = \Omega$	
	$A \cap \emptyset = \emptyset$	$A \cap \Omega = A$	
Idempotent laws:	$A \cup A = A$	$A \cap A = A$	
Complement laws:	$A \cup \bar{A} = \Omega$	$A \cap \bar{A} = \emptyset$	$\overline{(\bar{A})} = A$
Commutative laws:	$A \cap B = B \cap A$	$A \cup B = B \cup A$	
Associative laws:	$A \cup (B \cup C) = (A \cup B) \cup C$		
	$A \cap (B \cap C) = (A \cap B) \cap C$		

Distributive laws: $A \cup (B \cap C) = (A \cup B) \cap (A \cup C)$

$A \cap (B \cup C) = (A \cap B) \cup (A \cap C)$

DeMorgan's laws: $\overline{(A \cup B)} = \bar{A} \cap \bar{B}$

$\overline{(A \cap B)} = \bar{A} \cup \bar{B}$

Try illustrating each of the above laws by means of Venn diagrams! Can you see the analogy between the algebra of sets and the basic rules of arithmetic?

Appendix C

Mathematical Induction as a Method of Proof

As we know, many mathematical theorems are proved by using a particular pattern or procedure. For instance, one commonly used technique is the "counter example." Another general method of proof is *mathematical induction*. This technique is useful for proving a theorem which involves a variable which takes on only integral values. For instance, if we add the first two natural numbers, 1 and 2, we see that their sum is 3. Similarly, the first three--1, 2 and 3--have a sum of 6. Can we detect any pattern in this relationship? Suppose we realize from intuition (or past knowledge) that the sum of the first n natural numbers is $n(n + 1)/2$. We can prove this result using mathematical induction.

Proof by induction is based on the following underlying principle: Suppose a theorem concerns a positive integer n. We are able to establish that the theorem is true for $n = 1$. Then by assuming that the theorem holds true for some particular value of n, say $n = k$, we can show that the statement is true for the next value of n, that is, $n = k + 1$. We could then conclude that we can always proceed to the next value of n, and hence the theorem is true for all positive integral values of n.

In general, proof by induction involves three parts:

Verification: Direct evaluation of the theorem for one or more particular values of the integral variable n, usually $n = 1$.

Extension: Prove that the theorem is true for the value $n = k + 1$ under the assumption that it is true for $n = k$.

Conclusion: A combination of the verification and extension shows that the theorem is true for any positive integral values of n.

Examples of Proof by Induction

Example C-1: Show by inducation that $\sum_{i=1}^{n} i = \frac{n(n + 1)}{2}$.

Solution: In Example A-3, we established that this is true by using the result obtained in Example A-2. Now let us prove this result by induction. Note that in order to use induction we must have both sides of the equation provided. Following the procedure set out above:

Verification: For $n = 1$, we have $1 = 1(1 + 1)/2$. For $n = 2$, we have $1 + 2 = 2(3)/2$. In both cases we see that the r.h.s. and l.h.s. are equal.

Extension: Let us assume the statement is true for $n = k$; that is,

$$\sum_{i=1}^{k} i = \frac{k(k + 1)}{2}.$$

Based on this assumption, we will then show that it is true for $n = k + 1$; that is,

$$\sum_{i=1}^{k+1} i = \frac{(k + 1)(k + 2)}{2}.$$

Note that $\sum_{i=1}^{k+1} i$ can be written as $\sum_{i=1}^{k} i + (k + 1)$.

By the assumption $\sum_{i=1}^{k} i = \frac{k(k + 1)}{2}$; hence, we can write

$$\sum_{i=1}^{k+1} i = \frac{k(k + 1)}{2} + (k + 1) = (k + 1)[\frac{k}{2} + 1] = \frac{(k + 1)(k + 2)}{2},$$

which is what we wished to show.

Conclusion: We can then say that this statement is true for all positive integral values of n.

Example C-2: Show by induction that $n^3 + 2n$ is divisible by 3.
Solution: We must show that this expression $(n^3 + 2n)$ is a multiple of 3 regardless of the value of n.
Verification: For n = 1, we have

$$n^3 + 2n = 3$$

and for n = 2

$$n^3 + 2n = 12$$

which are both divisible by 3.
Extension: Now assume that it is true for n = k; that is

$$k^3 + 2k$$

is a multiple of 3. If n = k + 1, we have

$$\begin{aligned}(k + 1)^3 + 2(k + 1) &= k^3 + 3k^2 + 5k + 3\\ &= k^3 + 2k + 3k^2 + 3k + 3\\ &= (k^3 + 2k) + 3(k^2 + k + 1).\end{aligned}$$

By the assumption $k^3 + 2k$ is a multiple of 3 and clearly $3(k^2 + k + 1)$ is also a multiple of 3. Thus

$$(k + 1)^3 + 2(k + 1)$$

is a multiple of 3.
Conclusion: This expression is divisible by 3 for all positive integral values of n.

Appendix D

Binomial Expansions

Consider the expansions in positive, integral powers of the binomial $(a + b)$; that is,

$$(a + b)^2 = a^2 + 2ab + b^2$$

$$(a + b)^3 = a^3 + 3a^2b + 3ab^2 + b^3$$

$$(a + b)^4 = a^4 + 4a^3b + 6a^2b^2 + 4ab^3 + b^4$$

$$\vdots$$

$$(a + b)^n = a^n + na^{n-1}b + \frac{n(n - 1)}{2!} a^{n-2}b^2 + \cdots + nab^{n-1} + b^n.$$

What are the general properties of these expansions?

1. There are $(n + 1)$ terms in each expansion.
2. The sum of the exponents for a and b in each term is n.
3. The exponents of a decrease from n to 0 while those of b increase from 0 to n.
4. Coefficients of terms equidistant from the ends are the same.

How can we express the coefficient of the general term of this expression? The following theorem demonstrates how to determine the general coefficient.

Theorem D·1: *If* n *is a positive integer, then*

$$(a + b)^n = \sum_{j=0}^{n} \binom{n}{j} a^{n-j} b^j.$$

Proof:

This thoerem will be proved by induction.

Verification: For $n = 1$ we have $(a + b)^1 = a + b$ and for $n = 2$, $(a + b)^2 = a^2 + 2ab + b^2$. Both of which we know to be true.

Extension: Now assuming that the statment is true for $n = k$:

$$(a + b)^k = \sum_{j=0}^{k} \binom{k}{j} a^{k-j} b^j, \tag{1}$$

show that it is true for $n = k + 1$:

$$(a + b)^{k+1} = \sum_{j=0}^{k+1} \binom{k+1}{j} a^{k+1-j} b^j. \tag{2}$$

Since $(a + b)^{k+1} = (a + b)(a + b)^k$, we can substitute (1) for $(a + b)^k$ and obtain

$$(a + b)^{k+1} = (a + b) \sum_{j=0}^{k} \binom{k}{j} a^{k-j} b^j$$

$$= \sum_{j=0}^{k} \binom{k}{j} a^{k+1-j} b^j + \sum_{j=0}^{k} \binom{k}{j} a^{k-j} b^{j+1}$$

$$= a^{k+1} + \sum_{j=1}^{k} \binom{k}{j} a^{k+1-j} b^j + \sum_{j=0}^{k-1} \binom{k}{j} a^{k-j} b^{j+1} + b^{k+1}.$$

In the second summation, let $t = j + 1$. Then

$$(a + b)^{k+1} = a^{k+1} + \sum_{j=1}^{k} \binom{k}{j} a^{k+1-j} b^j + \sum_{t=1}^{k} \binom{k}{t-1} a^{k+1-t} b^t + b^{k+1}$$

$$= a^{k+1} + \sum_{j=1}^{k} \left[\binom{k}{j} + \binom{k}{j-1}\right] a^{k+1-j} b^j + b^{k+1}.$$

Since

$$\binom{k}{j} + \binom{k}{j-1} = \binom{k+1}{j}, \tag{3}$$

$$(a+b)^{k+1} = a^{k+1} + \sum_{j=1}^{k} \binom{k+1}{j} a^{k+1-j} b^j + b^{k+1}$$

$$= \sum_{j=0}^{k+1} \binom{k+1}{j} a^{k+1-j} b^j,$$

which is the expression given in (2).

Conclusion: Thus the theorem is true for all positive integral n.

Alternatively, the form of the coefficients for Theorem D·1 can also be obtained by a combinatoric argument. Ignoring the coefficient, it is quite clear that each term in the expansion of $(a + b)^n$ has the general pattern

$$a^{n-r} b^r$$

for $r = 0, 1, \ldots, n$. How did we obtain this term? The product $(a + b)^n$ contains the factor $(a + b)$ n times. The general term $a^{n-r}b^r$ is formed by selecting b from r factors and a from the remaining $n - r$. In how many ways can this selection be made? We are interested in the number of arrangements of r b's and $(n - r)$ a's:

$$\underbrace{b \ldots b}_{r} \quad \underbrace{a \ldots a}_{n-r}$$

Applying Theorem 3·3, we see that this can be done in

$$\frac{n!}{r!(n-r)!} = \binom{n}{r}$$

ways. □

The combination of the coefficients given in (3) is often referred to as <u>Pascal's rule</u>:

$$\binom{k+1}{r} = \binom{k}{r-1} + \binom{k}{r} \qquad \text{for } 1 \le r \le k.$$

This identity is useful for constructing tables of binomial coefficients. If, for example, we know the coefficients for n = 10, we can determine those for n = 11:

$$\binom{11}{r} = \binom{10}{r-1} + \binom{10}{r}.$$

Hence, from the coefficients for n = 2 we can successively build those for larger n.

Several useful identities arise as variants of Theorem D·1:

(i) Let a = 1 and b = t; then

$$(1 + t)^n = \sum_{i=0}^{n} \binom{n}{i} t^i.$$

(ii) Let a = 1 and b = 1; then

$$\sum_{i=0}^{n} \binom{n}{i} = 2^n.$$

(iii) Let a = p and b = 1 - p; then

$$\sum_{i=0}^{n} \binom{n}{i} p^{n-i} (1 - p)^i = 1.$$

Theorem D·2: *This identity is frequently referred to as Vandermonde's identity:*

$$\sum_{h=0}^{k} \binom{n}{h} \binom{m}{k-h} = \binom{n+m}{k}.$$

Proof:

From (i) under the special identities,

$$(1 + t)^{n+m} = \sum_{k=0}^{n+m} \binom{n+m}{k} t^k. \qquad (1)$$

With $(1 + t)^{n+m} = (1 + t)^n(1 + t)^m$ and a repeated application of the above identity,

$$(1 + t)^{n+m} = \left[\sum_{j=0}^{n} \binom{n}{j} t^j\right]\left[\sum_{\ell=0}^{m} \binom{m}{\ell} t^\ell\right]. \tag{2}$$

Equating (1) and (2) gives

$$\left[\sum_{j=0}^{n} \binom{n}{j} t^j\right]\left[\sum_{\ell=0}^{m} \binom{m}{\ell} t^\ell\right] = \sum_{k=0}^{n+m} \binom{n + m}{k} t^k. \tag{3}$$

Expanding the l.h.s., we have

$$\left[1 + nt + \frac{n(n - 1)}{2} t^2 + \cdots + \binom{n}{j} t^j + \cdots + t^n\right]\left[1 + mt + \frac{m(m - 1)}{2} t^2 + \cdots + \binom{m}{\ell} t^\ell + \cdots + t^m\right]. \tag{4}$$

If we take the product of these two finite series, what is the coefficient of the general term, say t^k? Clearly this exponent of t ranges from 0 to n + m, but how does its coefficient arise?

Multiplying the two expressions in (4), we see that

Exponent in Product Series	*First Series*	*Second Series*	*Product Series*
0	(1)	(1)	$(1)t^0$
1	nt (1)	(1) mt	$(n + m)t^1$
⋮	⋮	⋮	⋮
k	(1) $\binom{n}{1} t$ ⋮ $\binom{n}{n} t^n$	$\binom{m}{k} t^k$ $\binom{m}{k - 1} t^{k-1}$ ⋮ $\binom{m}{k - n} t^{k-n}$	$\left[\sum_{h=0}^{k} \binom{n}{h}\binom{m}{k - h}\right] t^k$
⋮	⋮	⋮	⋮
n + m	$(1)t^n$	$(1)t^m$	$(1)t^{n+m}$

Therefore, (4) can be written as

$$\sum_{k=0}^{n+m} t^k \left[\sum_{h=0}^{k} \binom{n}{h} \binom{m}{k-h} \right]. \tag{5}$$

Substituting (5) for the l.h.s. of (3) and equating coefficients of t^k, we have

$$\sum_{h=0}^{k} \binom{n}{h} \binom{m}{k-h} = \binom{n+m}{k}. \qquad \square$$

Appendix E

Infinite Series

E-1 The Limit of a Sequence

Consider an unending sequence $a_1, a_2, \ldots, a_n, \ldots$. This implies that there is a well-defined rule which orders the terms as first number, second number and so on. What do we mean by the limit of this sequence? To say that the limit of this sequence exists and equals a,

$$\lim_{n\to\infty} a_n = a,$$

means that as we make n larger and larger, the number a_n gets closer to a. More precisely,

$$\lim_{n\to\infty} a_n = a$$

if, corresponding to any small positive number h, it is possible to find a number m such that

$$|a_n - a| < h \qquad \text{for } n \geq m.$$

In general terms, given any measure of closeness, say h, and if we go far enough in the sequence, say beyond m, all a_n will differ from the limit a by a quantity smaller than h.

Example E-1: Consider the sequence 2.1, 2.01, 2.001, ..., with the general term

$$b_n = 2 + \left(\frac{1}{10}\right)^n.$$

Then

$$\lim_{n\to\infty} b_n = 2.$$

If we take n sufficiently large, we can make the difference between 2 and $2 + \left(\frac{1}{10}\right)^n$ less than any specified number. For example, consider n = 10 and n = 11. What is the difference between $2 + \left(\frac{1}{10}\right)^n$ and 2 in these cases?

Example E-2: In contrast to Example E-1, we see that the sequence 1, 0, 1, 0, ..., does not have a limit. There is no number b which the sequence is approaching.

E-2 Convergence of an Infinite Series

Let $u_1, u_2, \ldots, u_n, \ldots,$ be an unending sequence. The expression

$$u_1 + u_2 + u_3 + \cdots = \sum_{i=1}^{\infty} u_i$$

is called an *infinite series*. What is meant by an infinite sum? Of course, we know that it is not possible to add infinitely many numbers; thus, it is impossible to give the series a meaning simply by adding its terms. Denote by S_n the sum of the first n terms of the series; that is,

$$\begin{aligned} S_1 &= u_1 \\ S_2 &= u_1 + u_2 \\ &\vdots \\ S_n &= u_1 + u_2 + u_3 + \cdots + u_n \\ &\vdots \end{aligned}$$

In this way we have constructed a new unending <u>sequence</u>

$$S_1, S_2, \ldots, S_n, \ldots.$$

We refer to S_n as the nth partial sum. If as n becomes infinite, the nth partial sum approaches a limit, that is, if $\lim_{n\to\infty} S_n$ exists, then the series is said to *converge* and its limit S is called the *sum* of the series. Thus

$$u_1 + u_2 + \cdots = \lim_{n\to\infty} S_n = S.$$

If the limit does not exist, then the series is said to *diverge* and we can*not* assign any value to its sum.

E-3 Special Series

Geometric series: $1 + r + r^2 + \cdots$

If $|r| < 1$, this series is convergent with sum $1/(1 - r)$.

Power series: $1 + mx + \frac{m(m - 1)}{2!} x^2 + \frac{m(m - 1)(m - 2)}{3!} x^3 + \cdots$

If m is a positive integer, we saw in Theorem D·1 that this series reduces to a finite sum of (m + 1) terms. In this case there is no question of convergence. If, however, m is not a positive integer, then the series is infinite. It can be shown that the series

$$\text{converges if } |x| < 1$$

and

$$\text{diverges if } |x| \geq 1.$$

When the series converges, its sum is $(1 + x)^m$.

Negative binomial series: If m (in the power series) is a negative integer and $|t| < 1$, we have

$$1 + nt + \frac{n(n + 1)}{2!} t^2 + \cdots = \sum_{j=0}^{\infty} \binom{n + j - 1}{j} t^j = (1 - t)^{-n}.$$

It may be useful to note that this is a power series with $m = -n$ and $x = -t$.

Since this is a frequently encountered series, some particular cases may be of intetest:

(i) For $n = 1$,

$$1 + t + t^2 + \cdots = (1 - t)^{-1} = \frac{1}{1 - t}, \text{ the geometric series.}$$

(ii) For $n = 2$,

$$1 + 2t + 3t^2 + \cdots = (1 - t)^{-2} = \frac{1}{(1 - t)^2}.$$

(iii) For $n = 3$,

$$1 + 3t + 6t^2 + \cdots = (1 - t)^{-3} = \frac{1}{(1 - t)^3}.$$

Exponential series: $1 + \frac{1}{1!} + \frac{1}{2!} + \frac{1}{3!} + \cdots$

This is a convergent infinite series represented by the symbol e. To five decimal places, its value is 2.71828. More formally, we have

$$\lim_{n\to\infty} \left(1 + \frac{1}{n}\right)^n = e.$$

Also, for any real values x, we have the series

$$1 + \frac{x}{1!} + \frac{x^2}{2!} + \cdots$$

which converges to e^x or $\exp(x)$; that is,

$$\lim_{n\to\infty} \left(1 + \frac{x}{n}\right)^n = \exp(x).$$

More generally, we can write

$$1 + \frac{ax}{1!} + \frac{a^2x^2}{2!} + \frac{a^3x^3}{3!} + \cdots = \sum_{j=0}^{\infty} \frac{(ax)^j}{j!} = \exp(ax) = e^{ax}$$

or

$$\lim_{n\to\infty}\left(1 + \frac{ax}{n}\right)^n = \exp(ax)$$

or

$$\lim_{n\to\infty}\left(1 + \frac{x}{n}\right)^{an} = \exp(ax).$$

Tables

Table 1. Individual Binomial Probabilities $b(x; n, p)$ for $n = 2(1)10(5)25$ and $p = .01, .05, .1, .2, .25, .3, 1/3, .4, .5, .6, 2/3, .7, .75, .8, .9, .95, .99$

Table 2. Individual Poisson Probabilities $p(x; \lambda)$ for $\lambda = .1(.1)1.0(.5)5, 6, 8, 10$

Table 1. Individual Binomial Probabilities b(x; n, p)

n	x	.01	.05	.1	.2	.25	.3	1/3	.4
2	0	.9801	.9025	.8100	.6400	.5625	.4900	.4444	.3600
	1	.0198	.0950	.1800	.3200	.3750	.4200	.4444	.4800
	2	.0001	.0025	.0100	.0400	.0625	.0900	.1111	.1600
3	0	.9703	.8574	.7290	.5120	.4219	.3430	.2963	.2160
	1	.0294	.1354	.2430	.3840	.4219	.4410	.4444	.4320
	2	.0003	.0071	.0270	.0960	.1406	.1890	.2222	.2880
	3		.0001	.0010	.0080	.0156	.0270	.0370	.0640
4	0	.9606	.8145	.6561	.4096	.3164	.2401	.1975	.1296
	1	.0388	.1715	.2916	.4096	.4219	.4116	.3951	.3456
	2	.0006	.0135	.0486	.1536	.2109	.2646	.2963	.3456
	3		.0005	.0036	.0256	.0469	.0756	.0988	.1536
	4			.0001	.0016	.0039	.0081	.0123	.0256
5	0	.9510	.7738	.5905	.3277	.2373	.1681	.1317	.0778
	1	.0480	.2036	.3280	.4096	.3955	.3602	.3292	.2592
	2	.0010	.0214	.0729	.2048	.2637	.3087	.3292	.3456
	3		.0011	.0081	.0512	.0879	.1323	.1646	.2304
	4			.0005	.0064	.0146	.0284	.0412	.0768
	5				.0003	.0010	.0024	.0041	.0102
6	0	.9415	.7351	.5314	.2621	.1780	.1176	.0878	.0467
	1	.0571	.2321	.3543	.3932	.3560	.3025	.2634	.1866
	2	.0014	.0305	.0984	.2458	.2966	.3241	.3292	.3110
	3		.0021	.0146	.0819	.1318	.1852	.2195	.2765
	4		.0001	.0012	.0154	.0330	.0595	.0823	.1382
	5			.0001	.0015	.0044	.0102	.0165	.0369
	6				.0001	.0002	.0007	.0014	.0041
7	0	.9321	.6983	.4783	.2097	.1335	.0824	.0585	.0280
	1	.0659	.2573	.3720	.3670	.3115	.2471	.2048	.1306
	2	.0020	.0406	.1240	.2753	.3115	.3176	.3073	.2613
	3		.0036	.0230	.1147	.1730	.2269	.2561	.2903
	4		.0002	.0026	.0287	.0577	.0972	.1280	.1935
	5			.0002	.0043	.0115	.0250	.0384	.0774
	6				.0004	.0013	.0036	.0064	.0172
	7					.0001	.0002	.0005	.0016

(Table 1 continued)

p								
.5	.6	2/3	.7	.75	.8	.9	.95	.99
.2500	.1600	.1111	.0900	.0625	.0400	.0100	.0025	.0001
.5000	.4800	.4444	.4200	.3750	.3200	.1800	.0950	.0198
.2500	.3600	.4444	.4900	.5625	.6400	.8100	.9025	.9801
.1250	.0640	.0370	.0270	.0156	.0080	.0010	.0001	
.3750	.2880	.2222	.1890	.1406	.0960	.0270	.0071	.0003
.3750	.4320	.4444	.4410	.4219	.3840	.2430	.1354	.0294
.1250	.2160	.2963	.3430	.4219	.5120	.7290	.8574	.9703
.0625	.0256	.0123	.0081	.0039	.0016	.0001		
.2500	.1536	.0988	.0756	.0469	.0256	.0036	.0005	
.3750	.3456	.2963	.2646	.2109	.1536	.0486	.0135	.0006
.2500	.3456	.3951	.4116	.4219	.4096	.2916	.1715	.0388
.0625	.1296	.1975	.2401	.3164	.4096	.6561	.8145	.9606
.0312	.0102	.0041	.0024	.0010	.0003			
.1562	.0768	.0412	.0284	.0146	.0064	.0005		
.3125	.2304	.1646	.1323	.0879	.0512	.0081	.0011	
.3125	.3456	.3292	.3087	.2637	.2048	.0729	.0214	.0010
.1562	.2592	.3292	.3602	.3955	.4096	.3280	.2036	.0480
.0312	.0778	.1317	.1681	.2373	.3277	.5905	.7738	.9510
.0156	.0041	.0014	.0007	.0002	.0001			
.0938	.0369	.0165	.0102	.0044	.0015	.0001		
.2344	.1382	.0823	.0595	.0330	.0154	.0012	.0001	
.3125	.2765	.2195	.1852	.1318	.0819	.0146	.0021	
.2344	.3110	.3292	.3241	.2966	.2458	.0984	.0305	.0014
.0938	.1866	.2634	.3025	.3560	.3932	.3543	.2321	.0571
.0156	.0467	.0878	.1176	.1780	.2621	.5314	.7351	.9415
.0078	.0016	.0005	.0002	.0001				
.0547	.0172	.0064	.0036	.0013	.0004			
.1641	.0774	.0384	.0250	.0115	.0043	.0002		
.2734	.1935	.1280	.0972	.0577	.0287	.0026	.0002	
.2734	.2903	.2561	.2269	.1730	.1147	.0203	.0036	
.1641	.2613	.3073	.3176	.3115	.2753	.1240	.0406	.0020
.0547	.1306	.2048	.2471	.3115	.3670	.3720	.2573	.0659
.0078	.0280	.0585	.0824	.1335	.2097	.4783	.6983	.9321

(Table 1 continued)

n	x	.01	.05	.1	.2	.25	.3	1/3	.4
8	0	.9227	.6634	.4305	.1678	.1001	.0576	.0390	.0168
	1	.0746	.2793	.3826	.3355	.2670	.1977	.1561	.0896
	2	.0026	.0515	.1488	.2936	.3115	.2965	.2731	.2090
	3	.0001	.0054	.0331	.1468	.2076	.2541	.2731	.2787
	4		.0004	.0046	.0459	.0865	.1361	.1707	.2322
	5			.0004	.0092	.0231	.0467	.0683	.1239
	6				.0011	.0038	.0100	.0171	.0413
	7				.0001	.0004	.0012	.0024	.0079
	8						.0001	.0002	.0007
9	0	.9135	.6302	.3874	.1342	.0751	.0404	.0260	.0101
	1	.0830	.2985	.3874	.3020	.2253	.1556	.1171	.0605
	2	.0034	.0629	.1722	.3020	.3003	.2668	.2341	.1612
	3	.0001	.0077	.0446	.1762	.2336	.2668	.2731	.2508
	4		.0006	.0074	.0661	.1168	.1715	.2048	.2508
	5			.0008	.0165	.0389	.0735	.1024	.1672
	6			.0001	.0028	.0087	.0210	.0341	.0743
	7				.0003	.0012	.0039	.0073	.0212
	8					.0001	.0004	.0009	.0035
	9							.0001	.0003
10	0	.9044	.5987	.3487	.1074	.0563	.0282	.0173	.0060
	1	.0914	.3151	.3874	.2684	.1877	.1211	.0868	.0403
	2	.0042	.0746	.1937	.3020	.2816	.2335	.1951	.1209
	3	.0001	.0105	.0574	.2013	.2503	.2668	.2601	.2150
	4		.0010	.0112	.0881	.1460	.2001	.2276	.2508
	5		.0001	.0015	.0264	.0584	.1029	.1366	.2007
	6			.0001	.0055	.0162	.0368	.0569	.1115
	7				.0008	.0031	.0090	.0163	.0425
	8				.0001	.0004	.0014	.0030	.0106
	9						.0001	.0003	.0016
	10								.0001

(Table 1 continued)

p								
.5	.6	2/3	.7	.75	.8	.9	.95	.99
.0039	.0007	.0002	.0001					
.0312	.0079	.0024	.0012	.0004	.0001			
.1094	.0413	.0171	.0100	.0038	.0011			
.2188	.1239	.0683	.0467	.0231	.0092	.0004		
.2734	.2322	.1707	.1361	.0865	.0459	.0046	.0004	
.2188	.2787	.2731	.2541	.2076	.1468	.0331	.0054	.0001
.1094	.2090	.2731	.2965	.3115	.2936	.1488	.0515	.0026
.0312	.0896	.1561	.1977	.2670	.3355	.3826	.2793	.0746
.0039	.0168	.0390	.0576	.1001	.1678	.4305	.6634	.9227
.0020	.0003	.0001						
.0176	.0035	.0009	.0004	.0001				
.0703	.0212	.0073	.0039	.0012	.0003			
.1641	.0743	.0341	.0210	.0087	.0028	.0001		
.2461	.1672	.1024	.0735	.0389	.0165	.0008		
.2461	.2508	.2048	.1715	.1168	.0661	.0074	.0006	
.1641	.2508	.2731	.2668	.2336	.1762	.0446	.0077	.0001
.0703	.1612	.2341	.2668	.3003	.3020	.1722	.0629	.0034
.0176	.0605	.1171	.1556	.2253	.3020	.3874	.2985	.0830
.0020	.0101	.0260	.0404	.0751	.1342	.3874	.6302	.9135
.0010	.0001							
.0098	.0016	.0003	.0001					
.0439	.0106	.0030	.0014	.0004	.0001			
.1172	.0425	.0163	.0090	.0031	.0008			
.2051	.1115	.0569	.0368	.0162	.0055	.0001		
.2461	.2007	.1366	.1029	.0584	.0264	.0015	.0001	
.2051	.2508	.2276	.2001	.1460	.0881	.0112	.0010	
.1172	.2150	.2601	.2668	.2503	.2013	.0574	.0105	.0001
.0439	.1209	.1951	.2335	.2816	.3020	.1937	.0746	.0042
.0098	.0403	.0867	.1211	.1877	.2684	.3874	.3151	.0914
.0010	.0060	.0173	.0282	.0563	.1074	.3487	.5987	.9044

(Table 1 continued)

n	x	.01	.05	.1	.2	.25	.3	1/3	.4
					p				
15	0	.8601	.4633	.2059	.0352	.0134	.0047	.0023	.0005
	1	.1303	.3658	.3432	.1319	.0668	.0305	.0171	.0047
	2	.0092	.1348	.2669	.2309	.1559	.0916	.0599	.0219
	3	.0004	.0307	.1285	.2501	.2252	.1700	.1299	.0634
	4		.0049	.0428	.1876	.2252	.2186	.1948	.1268
	5		.0006	.0105	.1032	.1651	.2061	.2143	.1859
	6			.0019	.0430	.0917	.1472	.1786	.2066
	7			.0003	.0138	.0393	.0811	.1148	.1771
	8				.0035	.0131	.0348	.0574	.1181
	9				.0007	.0034	.0116	.0223	.0612
	10				.0001	.0007	.0030	.0067	.0245
	11					.0001	.0006	.0015	.0074
	12						.0001	.0003	.0016
	13								.0003
	14								
	15								
20	0	.8179	.3585	.1216	.0115	.0032	.0008	.0003	
	1	.1652	.3774	.2702	.0577	.0211	.0068	.0030	.0005
	2	.0159	.1887	.2852	.1369	.0669	.0278	.0143	.0031
	3	.0010	.0596	.1901	.2054	.1339	.0716	.0429	.0124
	4		.0133	.0898	.2182	.1897	.1304	.0911	.0350
	5		.0022	.0319	.1746	.2023	.1789	.1457	.0746
	6		.0003	.0089	.1091	.1686	.1916	.1821	.1244
	7			.0020	.0546	.1124	.1643	.1821	.1659
	8			.0004	.0222	.0609	.1144	.1480	.1797
	9				.0074	.0271	.0654	.0987	.1597
	10				.0020	.0099	.0308	.0543	.1171
	11				.0005	.0030	.0120	.0247	.0710
	12				.0001	.0008	.0039	.0092	.0355
	13					.0002	.0010	.0028	.0146
	14						.0002	.0007	.0048
	15							.0001	.0013
	16								.0003
	17								
	18								
	19								
	20								

(Table 1 continued)

p								
.5	.6	2/3	.7	.75	.8	.9	.95	.99
.0005								
.0032	.0003							
.0139	.0016	.0003	.0001					
.0417	.0074	.0015	.0006	.0001				
.0916	.0245	.0067	.0030	.0007	.0001			
.1527	.0612	.0223	.0116	.0034	.0007			
.1964	.1181	.0574	.0348	.0131	.0035			
.1964	.1771	.1148	.0811	.0393	.0138	.0003		
.1527	.2066	.1786	.1472	.0917	.0430	.0019		
.0916	.1859	.2143	.2061	.1651	.1032	.0105	.0006	
.0417	.1268	.1948	.2186	.2252	.1876	.0428	.0049	
.0139	.0634	.1299	.1700	.2252	.2501	.1285	.0307	.0004
.0032	.0219	.0599	.0916	.1559	.2309	.2669	.1348	.0092
.0005	.0047	.0171	.0305	.0668	.1319	.3432	.3653	.1303
	.0005	.0023	.0047	.0134	.0352	.2059	.4633	.8601

.5	.6	2/3	.7	.75	.8	.9	.95	.99
.0002								
.0011								
.0046	.0003							
.0148	.0013	.0001						
.0370	.0048	.0007	.0002					
.0739	.0146	.0028	.0010					
.1201	.0355	.0092	.0039	.0008	.0001			
.1602	.0710	.0247	.0120	.0030	.0005			
.1762	.1171	.0543	.0308	.0099	.0020			
.1602	.1597	.0987	.0654	.0271	.0074			
.1201	.1797	.1480	.1144	.0609	.0222	.0004		
.0739	.1659	.1821	.1643	.1124	.0546	.0020		
.0370	.1244	.1821	.1916	.1686	.1091	.0089	.0003	
.0148	.0746	.1457	.1789	.2023	.1746	.0319	.0022	
.0046	.0350	.0911	.1304	.1897	.2182	.0898	.0133	
.0011	.0124	.0429	.0716	.1339	.2054	.1901	.0596	.0010
.0002	.0031	.0143	.0278	.0669	.1369	.2852	.1887	.0159
	.0005	.0030	.0068	.0211	.0577	.2072	.3774	.1652
		.0003	.0008	.0032	.0115	.1216	.3585	.8179

(Table 1 continued)

n	x	p .01	.05	.1	.2	.25	.3	2/3	.4
25	0	.7778	.2774	.0718	.0038	.0008	.0001		
	1	.1964	.3650	.1994	.0236	.0063	.0014	.0005	
	2	.0238	.2305	.2659	.0708	.0251	.0074	.0030	.0004
	3	.0018	.0930	.2265	.1358	.0641	.0243	.0114	.0019
	4	.0001	.0269	.1384	.1867	.1175	.0572	.0313	.0071
	5		.0060	.0646	.1960	.1645	.1030	.0658	.0199
	6		.0010	.0239	.1634	.1828	.1472	.1096	.0442
	7		.0002	.0072	.1108	.1654	.1712	.1487	.0800
	8			.0018	.0624	.1241	.1651	.1673	.1200
	9			.0004	.0294	.0781	.1336	.1580	.1511
	10			.0001	.0118	.0417	.0916	.1264	.1612
	11				.0040	.0189	.0536	.0862	.1465
	12				.0012	.0074	.0268	.0503	.1140
	13				.0003	.0025	.0115	.0251	.0760
	14				.0001	.0007	.0042	.0108	.0434
	15					.0002	.0013	.0040	.0212
	16						.0004	.0012	.0088
	17						.0001	.0003	.0031
	18							.0001	.0009
	19								.0002
	20								
	21								
	22								
	23								
	24								
	25								

(Table 1 continued)

				p				
.5	.6	2/3	.7	.75	.8	.9	.95	.99
.0001								
.0004								
.0016								
.0053	.0002							
.0143	.0009	.0001						
.0322	.0031	.0003	.0001					
.0609	.0088	.0012	.0004					
.0974	.0212	.0040	.0013	.0002				
.1328	.0434	.0108	.0042	.0007	.0001			
.1550	.0760	.0251	.0115	.0025	.0003			
.1550	.1140	.0503	.0268	.0074	.0012			
.1328	.1465	.0862	.0536	.0189	.0040			
.0974	.1612	.1264	.0916	.0417	.0118	.0001		
.0609	.1511	.1580	.1336	.0781	.0294	.0004		
.0322	.1200	.1673	.1651	.1241	.0624	.0018		
.0143	.0800	.1487	.1712	.1654	.1108	.0072	.0002	
.0053	.0442	.1096	.1472	.1828	.1634	.0239	.0010	
.0016	.0199	.0658	.1030	.1645	.1960	.0646	.0060	
.0004	.0071	.0313	.0572	.1175	.1867	.1384	.0269	.0001
.0001	.0019	.0114	.0243	.0641	.1358	.2265	.0930	.0018
	.0004	.0030	.0074	.0251	.0708	.2659	.2305	.0238
		.0005	.0014	.0063	.0236	.1994	.3650	.1964
			.0001	.0008	.0038	.0718	.2774	.7778

Table 2. Individual Poisson Probabilities $p(x; \lambda)$

x	λ = .1	.2	.3	.4	.5	.6	.7
0	.9048	.8187	.7408	.6703	.6065	.5488	.4966
1	.0905	.1637	.2222	.2681	.3033	.3293	.3476
2	.0045	.0164	.0333	.0536	.0758	.0988	.1217
3	.0002	.0011	.0033	.0072	.0126	.0198	.0284
4		.0001	.0003	.0007	.0016	.0030	.0050
5				.0001	.0002	.0004	.0007
6							.0001

x	λ = .8	.9	1.0	1.5	2.0	2.5	3.0
0	.4493	.4066	.3679	.2231	.1353	.0821	.0498
1	.3595	.3659	.3679	.3347	.2707	.2052	.1494
2	.1438	.1647	.1839	.2510	.2707	.2565	.2240
3	.0383	.0494	.0613	.1255	.1804	.2138	.2240
4	.0077	.0111	.0153	.0471	.0902	.1336	.1680
5	.0012	.0020	.0031	.0141	.0361	.0668	.1008
6	.0002	.0003	.0005	.0035	.0120	.0278	.0504
7			.0001	.0008	.0034	.0099	.0216
8				.0001	.0009	.0031	.0081
9					.0002	.0009	.0027
10						.0002	.0008
11							.0002
12							.0001
13							
14							
15							
16							
17							
18							
19							
20							
21							
22							
23							
24							

(Table 2 continued)

x	λ 3.5	4.0	4.5	5.0	6.0	8.0	10.0
0	.0302	.0183	.0111	.0067	.0025	.0003	.0000
1	.1057	.0733	.0500	.0337	.0149	.0027	.0005
2	.1850	.1465	.1125	.0842	.0446	.0107	.0023
3	.2158	.1954	.1687	.1404	.0892	.0286	.0076
4	.1888	.1954	.1898	.1755	.1339	.0572	.0189
5	.1322	.1563	.1708	.1755	.1606	.0916	.0378
6	.0771	.1042	.1281	.1462	.1606	.1221	.0631
7	.0385	.0595	.0824	.1044	.1377	.1396	.0901
8	.0169	.0298	.0463	.0653	.1033	.1396	.1126
9	.0066	.0132	.0232	.0363	.0688	.1241	.1251
10	.0023	.0053	.0104	.0181	.0413	.0993	.1251
11	.0007	.0019	.0043	.0082	.0225	.0722	.1137
12	.0002	.0006	.0016	.0034	.0113	.0481	.0948
13	.0001	.0002	.0006	.0013	.0052	.0296	.0729
14		.0001	.0002	.0005	.0022	.0169	.0521
15			.0001	.0002	.0009	.0090	.0347
16					.0003	.0045	.0217
17					.0001	.0021	.0128
18						.0009	.0071
19						.0004	.0037
20						.0002	.0019
21						.0001	.0009
22							.0004
23							.0002
24							.0001

Answers To Selected Problems

Chapter 2

1. (a) Yes, (b) No, $Pr(E_1) < 0$, (c) No, $\Sigma\ Pr(E_i) < 1$,
   (d) No, $\Sigma\ Pr(E_i) > 1$.

2. (a) $Pr(\bar{A}) = .45$, $Pr(\bar{C}) = .7$, $Pr(A \cup C) = .65$, $Pr(B \cup C) = .65$, $Pr(A \cup B \cup C) = 1$. (b) $Pr(A \cup B) = .90 = .55 + .35$, since A and B are m.e. events. (c) $Pr(A \cup C) = .65 \neq .55 + .30$, since A and C are <u>not</u> m.e. events.

3. No, not necessarily. This result would be true if the probability model is uniform.

4. (a) $Pr(\bar{A}) = 1/2$, $Pr(\bar{C}) = 2/3$, $Pr(A \cup C) = 2/3$, $Pr(B \cup C) = 2/3$, $Pr(A \cup B \cup C) = 1$.

5. (a) Consider the event "one boy." In S_1 the event "one boy" is composed of {BGG, GBG, GGB}. Hence, under S_1, Pr(one boy) = 3/8 while using S_2, Pr(one boy) = 1/4. These results are not consistent.
   (b) Observe a large number of families with three children and see which assignment of probability seems more reasonable.
   (c) Since from S_1 to S_2 is a many-to-one correspondence, assigning probabilities in S_1 uniquely determines the probabilities in S_2. In contrast S_2 to S_1 has a one-to-many correspondence. Hence for a given assignment in S_2, many different probability assignments in S_1 are possible.

6. (a) (i)

green \ red	1	2	3	4	5	6
1	(1, 1)	(1, 2)	(1, 3)	(1, 4)	(1, 5)	(1, 6)
2	(2, 1)	(2, 2)	.	.	.	(2, 6)
3	.					.
4	.					.
5	.					.
6	(6, 1)	.	.	.	.	(6, 6)

(ii) $S = \{(g, r) \mid g = 1, 2, \ldots, 6 \text{ and } r = 1, 2, \ldots, 6\}$

(b) $A = \{(1, 1)(2, 2)(3, 3)(4, 4)(5, 5)(6, 6)\}$ or $A = \{(g, r) \mid g = r \text{ for } g = 1, 2, 3, \ldots, 6 \text{ and } r = 1, 2, \ldots, 6\}$, $B = \{(g, r) \mid g \geq r + 2\}$, $C = \{(g, r) \mid g + r = 8\}$.

(c) (i) The number on the red die is twice that on green.

(ii) The sum is greater than 5.

(iii) The two outcomes are different.

(d) $\Pr(A) = 1/6$, $\Pr(B) = 5/18$, $\Pr(C) = 5/36$, $\Pr(r = 2g) = 1/12$, $\Pr(r + g > 5) = 13/18$, $\Pr(r \neq g) = 5/6$.

7. 18/36

8. (a) Hint: $B = (B \cap \bar{A}) \cup (B \cap A) = (B \cap \bar{A}) \cup A$

13. (a) $S = \{(x, y) \mid x = 0, 1, \ldots, 5 \text{ and } y = 0, 1, 2, 3\}$

(d) 3/8, 5/12, 1/6, 1/8, 11/24, 1/24, 1/2, (e) 11/24, 1/2

14. (a) 2/3, 1/3, 1/2, 1/2, 1/6, 5/6, 2/3, 1/3

17. .13, 0, .17, .55, .45, .15

18. (b) 1/2, 1/2, 1/5, 4/5, 1/2, 1/5, (c) No change.

19. (a) $S = \{R_1R_2, R_1W_2, R_1P_2, W_1R_2, W_1W_2, W_1P_2, P_1R_2, P_1W_2, P_1P_2\}$

(e) Each color has an equal representation.

(f) 5/9, 2/9, 8/9, 1/3, 4/9, 5/9, 8/9, 2/9, 5/9, 5/9

20.

	N	$\bar{N}$	
D	.002	.008	.01
$\bar{D}$	.398	.592	.99
	.4	.6	

21. (a) Let the event (A, D, B, C) represent A selecting A's turtle, B selecting D's turtle, C selecting B's turtle and D selecting C's turtle. Thus S consists of the 4! = 24 orderings of the letters A, B, C, D.
    (b) 1/24, 3/8, 1/4, 1/12

22. (c) 3/7, (d) 3/7

23. (a) 3/8, 7/8, 1/2, 1/2, 1/2, 1/2
    (b) Pr{no boy or more than one boy} = 5/8, Pr{no boy} = 1/8, Pr{at least two boys} = 1/2
    (c) (iv) and (vi)
    (d) Pr{exactly one boy} = 3/8, Pr{more boys than girls} = 1/2, Pr{only the eldest child is a boy} = 1/8
    (e) (iv) and (vi)

24. .8

26. (a) Yes, yes.
    (b) $Q = (Q \cap C_1) \cup (Q \cap C_2) \cup \cdots \cup (Q \cap C_{35})$. We would need $Pr(Q \cap C_i)$ for i = 1, 2, ..., 35.

27. (a) $S = \{OGP, OG\bar{P}, O\bar{G}P, O\bar{G}\bar{P}, \bar{O}GP, \bar{O}G\bar{P}, \bar{O}\bar{G}\bar{P}, \bar{O}\bar{G}P\}$ or $S = \{E_1, E_2, E_3, E_4, E_5, E_6, E_7, E_8\}$
    (d) $Pr(E_i) = 1/8$ for i = 1, 2, ..., 8

Chapter 3

1. (a) 3024, (b) 6!, (c) n(n - 1), (d) k

2. (a) 120, (b) 1, (c) m(m - 1)/2, (d) t, (e) 500(499)/2, (f) k

5. (a) 9, (b) 11, (c) 10
6. The number of subsets of size r is the same as the number of subsets of size (n - r).
7. (a) 22, 20, (b) 10/11
8. (b) 1/4, 1/2, 1/2
9. (b) 1/4, 1/2, 1/2
10. (a) 8!, (b) 1/28
11. 44
12. 4095
13. 8/17, 4/17, 1/17
14. 60 are needed
15. 190
16. 256, 1/2
17. 14!/5!3!3!3!
18. 8^4, $_8P_4$
19. 6/15, $1/\binom{15}{6}$
20. 10!/4!2!, 1/30, 1/5
21. 1/35, 1/144
22. 3071
23. 1/55, 1/66, 6/55
24. 1/126, 1/21, 4/9
25. 60, 1/5
26. 1/35, 1/35, 1/140
27. 1/6
28. $\binom{14}{2}/\binom{25}{2}$, $\binom{11}{2}/\binom{25}{2}$, $\binom{14}{1}\binom{11}{1}/\binom{25}{2}$

29. 77/300, which is not the same as the answer to 28 (c).

30. 2/15, 1/3, 2/3

31. 4/25, 9/25, 16/25 .

32. 12/49, 1/4, 1/25

33. $\binom{N - r + 1}{1} \Big/ \binom{N}{r} = (N - r + 1)r!/N!$

34. 1/100, 1/10

35. 30/49

36. ${}_N P_k / N^k$

37. (a) A typical point in S is the six-tuple $(x_1, x_2, \ldots, x_6)$ where x_i is R or B. In all S will have 64 points.
    (b) The uniform model is not appropriate since there are not an equal number of red and blue balls.

38. Typical point in S: $(x_1, x_2, \ldots, x_k)$ where x_i can be 1, 2, ..., 6. S contains 6^k points. Here the uniform model is appropriate.

39. d(d - 1)/k(k - 1)

40. There are 52!/4!48! points in S.
    (a) $\binom{46}{3} \Big/ \binom{52}{4}$, (b) $\binom{5}{3} \Big/ \binom{52}{4}$, (c) $\binom{52 - r}{3} \Big/ \binom{52}{4}$,
    (d) $1 - \left[\binom{52 - r}{4} \Big/ \binom{52}{4}\right]$

41. $\dfrac{13!}{4!3!2!2!}$

42. $\dfrac{3!10!}{2!3!3!4!} \div 3^{10}$

43. (a) A typical point in S: (W, M), where each of W and M can be filled in four ways. Hence, S contains 16 points. The uniform model can be applied.

(b) Pr(E) = 1/4 and Pr(F) = 1/4. $E \cap \bar{F}$ is the selection of any married couple other than Mr. and Mrs. A.

44. 3/14, 1/4, 1/6

45. 1/4

Chapter 4

1. (a) $Pr(A) = 7/10$, $Pr(E_1 \mid A) = 2/7$, $Pr(E_2 \mid A) = 0$, $Pr(E_3 \mid A) = 3/14$, $Pr(E_4 \mid A) = 1/2$, $Pr(E_5 \mid A) = 0$.

   (b) Since $0 \le Pr(E_i \mid A) \le 1$ for all i and $\sum_i Pr(E_i \mid A) = 1$, this is a proper probability model.

2. (a) $Pr\{(1, 3) \mid sum = 4\} = Pr\{(3, 1) \mid sum = 4\} = Pr\{(2, 2) \mid sum = 4\} = 1/3$ and all other points have zero probability.

   (b) 2/3, (c) 2/3

3. $Pr(E_i) = 1/k$ for i = 1, 2, ..., k. Let $A = E_m \cup E_{m+1} \cup \cdots \cup E_{m+r}$; hence $Pr(A) = (r + 1)/k$. Then $Pr(E_i \mid A) = 1/(r + 1)$ if $E_i \subset A$ and $Pr(E_i \mid A) = 0$ if $E_i \cap A = \phi$, which gives a uniform probability model with r + 1 events.

4. (a) 1/2, (b) 1/6

5. (a) 1/2, (b) 1/2

6. (a)

	A	$\bar{A}$
B	.1	.3
$\bar{B}$	.1	.5

   (b) (i) 1/4, (ii) 1/2, (iii) 5/6, (iv) 9/10

7. (c) Pr(three heads) = 1/8, Pr(three heads | at least two heads) = 1/4.

8. $Pr(A \mid B) = 0$; in the conditional sample space, A is an impossible event.

9. $\Pr(A \mid B) = 1$; in the conditional sample space, A is a certain event. If we know B has occurred, A must have occurred.

10. (b) 1/24, (c) 1/4, 1/2, 1/8, 5/8, 1/4, 1/2, 1/2, 3/4

11. $\Pr(A \cap B \cap C) = \Pr[(A \cap B) \cap C] = \Pr(A \cap B)\Pr(C \mid A \cap B) = \Pr(A)\Pr(B \mid A)\Pr(C \mid A \cap B)$. This is proved using successive applications of the multiplication law.

12. (a) $b/(w + b - 1)$, (b) $w/(w + b)$, (c) $b/(b + w - 1)$

13. (a) $b/(w + b)$, (b) $w/(w + b)$, (c) $b/(w + b)$

14. Let R_1 be an adverse reaction to first shot and R_2 be an adverse reaction to second shot. Then $\Pr(R_1) = .4$, $\Pr(\bar{R}_1) = .6$, $\Pr(R_2 \mid R_1) = 0$, $\Pr(R_2 \mid \bar{R}_1) = .6$
   (a) $\Pr(R_2 \cap \bar{R}_1) = \Pr(R_2 \mid \bar{R}_1)\Pr(\bar{R}_1) = .36$
   (b) $\Pr(R_1 \cup R_2) = .4 + .36 = .76$ since R_1 and R_2 are m.e.

15.

E_i	3, 0, 0	0, 3, 0	0, 0, 3	1, 2, 0	2, 1, 0	0, 2, 1
$\Pr(E_i)$	$\frac{1}{27}$	$\frac{1}{27}$	$\frac{1}{27}$	$\frac{3}{27}$	$\frac{3}{27}$	$\frac{3}{27}$

E_i	1, 0, 2	2, 0, 1	0, 1, 2	1, 1, 1
$\Pr(E_i)$	$\frac{3}{27}$	$\frac{3}{27}$	$\frac{3}{27}$	$\frac{6}{27}$

$\Pr(\text{all three in one box} \mid \text{at least two in same box}) = \frac{3}{27} \div \frac{21}{27} = \frac{1}{7}$

16. (a) $\Pr(LL) = .4$, $\Pr(LR) = .1$, $\Pr(RL) = .3$, $\Pr(RR) = .2$,
   (b) $\Pr(\text{Right on 2}) = \Pr(LR) + \Pr(RR) = .1 + .2 = .3$
   (c) $\Pr(\text{R on 1st} \mid \text{R on 2nd}) = .2/.3 = 2/3$

17. $$\frac{\binom{4}{2} + \binom{6}{2} + \binom{2}{2}}{\binom{12}{2}} = \frac{1}{3}$$

18. 1/32, 1/16, 1/8, 1/4, 1

19. $\binom{r}{2}\Big/\binom{N}{2}$, $\binom{r}{1}\binom{N - r}{1}\Big/\binom{N}{2}$

20. (b) 3/11, 1/3, 3/7

21. .7347

22. 1/15, 1/2, 16/17

23. 1/2, 44/91

24. .15, .15

25. (b) 5/8, (c) 3/8, (d) 7/10

28. 53/70, 17/70

30. (b) (i) 119/120, (ii) 5/12

31. 9/32

32. 1/8

33. 9/59

34. 3/32

35. $Pr(1st \mid S) = 5/11$, $Pr(2nd \mid S) = 3/11$

36. (a) 21/23, (b) .87

37. (a) 19/217, (b) Error = $(+ \cap \bar{D})$ or $(- \cap D)$,
    (c) Pr(error) = .0995

38. (a) 95/105, (c) .075

Chapter 5

2. (a) .0001, (b) .9999

3. (a), (b), (c) are independent

4. $Pr(G) = 1/4$, $Pr(G \mid H) = 1/4$; G and H are independent.

6. 9

7. 1/14, 2/7

8. 1/6, 1/4; events A and B not independent.

9. True, only if A and B are m.e.

10. .27, no, 27/40

11. $(.8)^3(.4)^3$

12. (a) $(a + b)/n$, $(a + c)/n$, $a/(a + c)$, $a/(a + b)$

    (b) If $ad - bc = 0$, then $b = ad/c$. Consider $Pr(M) = (a + b)/(a + b + c + d)$. Let $b = ad/c$ and show that $Pr(M) = a/(a + c)$, which is also $Pr(M \mid E)$; hence M and E are independent.

    (c) Apply Theorem 5·1.

13. (a) $(.8)^2(.5)^2$, (b) $(.8)^2(.5)^2 + 4(.8)(.2)(.5)^2 + (.2)^2(.5)^2$

    (c) $(.8)^2(.5)^2 + 4(.8)(.2)(.5)^2 + (.2)^2(.5)^2$

14. (a) $(.5)^4$, (b) $6(.5)^4$, (c) $6(.5)^4$

15. $Pr(E \cap F \mid F) = Pr(E)$; that is, proportion of outcomes in $E \cap F$ relative to F is the same as the proportion of outcomes in E relative to S. This is a result of the independence of E and F.

16. (a) $(.1)^6$, $(.9)^6$, $6(.1)(.9)^5$, $1 - (.9)^6$, $1 - (.9)^6 - 6(.1)(.9)^5$

    (b) Independence

17. p^n, $(1 - p)^n$, $np(1 - p)^{n-1}$, $1 - (1 - p)^n$, $\binom{n}{k}p^k(1 - p)^{n-k}$

18. (a) $p^2 + (1 - p)p^2$, (b) $p_1p_2 + (1 - p_1)p_2p_3$

19. (a) $Pr(R_2 \mid R_1) = 3/8$, $Pr(R_2) = Pr(R_1 \cap R_2) + Pr(W_1 \cap R_2)$
    $= 3/8 \cdot 3/8 + 5/8 \cdot 3/8 = 3/8$;

    therefore, R_1 and R_2 are independent.

    (b) $Pr(R_2 \mid R_1) = 2/7$, $Pr(R_2) = 3/8$; therefore R_1 and R_2 are not independent.

20. $1 - (1 - \theta)^k$

23. (a) $Pr(A) = 11/16$, $Pr(B) = 7/16$, $Pr(C) = 9/16$

    (b) $Pr(A \cap B) = 4/16 \neq [11/16][7/16]$; hence, A and B are not independent. A, B and C cannot be mutually independent.

    (c) $Pr(A \mid B) = 4/7 \neq Pr(A)$

    (d) 1/36, 2/36, 5/36, 10/36, 1/36, 5/36, 2/36, 10/36

25. $Pr(I) = 1/5$, $Pr(II) = 4/5$, $Pr(R \mid I) = 2/5$, $Pr(R \mid II) = 1/5$, $Pr(I \mid R) = 1/3$, $Pr(II \mid R) = 2/3$

26. (a) $\Pr(\mathrm{I} \mid n\mathrm{R}) = \dfrac{(1/5)(2/5)^n}{(2/5)^n(1/5) + 4/5(1/5)^n} = \dfrac{2^n}{2^n + 4}$

$\Pr(\mathrm{II} \mid n\mathrm{R}) = \dfrac{(4/5)(1/5)^n}{(2/5)^n 1/5 + 4/5(1/5)^n} = \dfrac{4}{2^n + 4}$

(b) $n = 2$

(c) As n gets large, $\Pr(\mathrm{I} \mid n\mathrm{R})$ approaches 1 and $\Pr(\mathrm{II} \mid n\mathrm{R})$ approaches zero; that is, if all n balls are red, it is nearly certain that they are being drawn from box I.

27. (a) $p_S p_T(2 - p_S p_T)$

(b) Rel(I) > Rel(II) is equivalent to $(1 - p_S)(1 - p_T) > 0$. Thus system I is always more reliable than II unless $p_S = 1$ or $p_T = 1$ or both p_S and p_T are equal to 1.

29. I: Aa × aa → all Aa

II: Aa × Aa → 1/4 AA, 1/2 Aa, 1/4 aa. Given an equal number of each type of mating:

(a) 1(1/2) + 1/2(1/2) = 3/4, (b) 2/3

30. (a) 2/3, (b) 1/2

Chapter 6

1. (a)

E_i	H	TH	TTH	TTT
$\Pr(E_i)$	$\frac{3}{8}$	$\left(\frac{5}{8}\right)\left(\frac{3}{8}\right)$	$\left(\frac{5}{8}\right)^2\left(\frac{3}{8}\right)$	$\left(\frac{5}{8}\right)^3$
x	1	2	3	3

Assumption: Independence

(b)

x	1	2	3
f(x)	3/8	15/64	25/64
F(x)	3/8	39/64	1

2. (b)

m	0	1	2	3
Pr(M = m)	1/3	1/2	0	1/6

(c) Pr(M = n) = 1/n!, Pr(M = n - 1) = 0

It is not possible to have (n - 1) matches, since (n - 1) matches implies n matches.

3. (a) 1/12, (b)

x	-1	1	3
w	-1	0	1
f(x)	1/4	1/4	1/2

4. (b)

x	2	3	4	5	6	7	8
f(x)	$\frac{1}{16}$	$\frac{2}{16}$	$\frac{3}{16}$	$\frac{4}{16}$	$\frac{3}{16}$	$\frac{2}{16}$	$\frac{1}{16}$

5. (b)

x	2	3	4	5	6	7	8
f(x)	$\frac{1}{100}$	$\frac{4}{100}$	$\frac{10}{100}$	$\frac{20}{100}$	$\frac{25}{100}$	$\frac{24}{100}$	$\frac{16}{100}$

6. (a)

x	1.5	3	5
f(x)	1/2	1/3	1/6

(b)

y	3	6	10
f(y)	1/2	1/3	1/6

7. (b)

x	4	5	6	7	8	9
f(x)	.1	.3	.3	0	.2	.1

(c) .7, .1, .3, .6

8. (a) 9/80, (c) 9/16

9. (a)

x	0	1	2	3
f(x)	1/8	3/8	3/8	1/8

(b)

y	-3	-1	1	3
g(y)	1/8	3/8	3/8	1/8

(c)

y \ x	0	1	2	3
-3	1/8	0	0	0
-1	0	3/8	0	0
1	0	0	3/8	0
3	0	0	0	1/8

(d) X and Y are not independent.

10. (a) $S = \{(x_1, x_2, x_3, x_4, x_5) \mid x_i = D \text{ or } N\}$. There will be $2^5 = 32$ points in S.

(b)

x	1	2	3	4	5
f(x)	1/16	4/16	6/16	4/16	1/16

11. (c) Using part (a):

x	0	1	2	3	4	5
f(x)	$\left(\frac{1}{2}\right)^5$	$5\left(\frac{1}{2}\right)^5$	$10\left(\frac{1}{2}\right)^5$	$10\left(\frac{1}{2}\right)^5$	$5\left(\frac{1}{2}\right)^5$	$\left(\frac{1}{2}\right)^5$

Using part (b):

x	0	1	2	3	4	5
g(x)	$\left(\frac{1}{3}\right)^5$	$5\left(\frac{2}{3}\right)\left(\frac{1}{3}\right)^4$	$10\left(\frac{2}{3}\right)^2\left(\frac{1}{3}\right)^3$	$10\left(\frac{2}{3}\right)^3\left(\frac{1}{3}\right)^2$	$5\left(\frac{2}{3}\right)^4\left(\frac{1}{3}\right)$	$\left(\frac{2}{3}\right)^5$

13. (a)

x	0	1	2	3
f(x)	2/6	3/6	0	1/6

(b) The p.f. of X remains the same since each ordering is equally likely.

14. (a)

s \ m	1	2	3	4	5	6
2	1/36	0	0	0	0	0
3	0	2/36	0	0	0	0
4	0	1/36	2/36	0	0	0
5	0	0	2/36	2/36	0	0
6	0	0	1/36	2/36	2/36	0

14. (a) continued

s \ m	1	2	3	4	5	6
7	0	0	0	2/36	2/36	2/36
8	0	0	0	1/36	2/36	2/36
9	0	0	0	0	2/36	2/36
10	0	0	0	0	1/36	2/36
11	0	0	0	0	0	2/36
12	0	0	0	0	0	1/36

(b)

m	1	2	3	4	5	6
g(m)	$\frac{1}{36}$	$\frac{3}{36}$	$\frac{5}{36}$	$\frac{7}{36}$	$\frac{9}{36}$	$\frac{11}{36}$

s	2	3	4	5	6	7	8	9	10	11	12
h(s)	$\frac{1}{36}$	$\frac{2}{36}$	$\frac{3}{36}$	$\frac{4}{36}$	$\frac{5}{36}$	$\frac{6}{36}$	$\frac{5}{36}$	$\frac{4}{36}$	$\frac{3}{36}$	$\frac{2}{36}$	$\frac{1}{36}$

(c) No, since $f(m, s) \neq g(m)h(s)$ for all m and s.

15. Sample space:

Boxes 1	2	3	Probability
4	0	0	1/81
0	4	0	1/81
0	0	4	1/81
3	1	0	4/81
1	3	0	4/81
0	1	3	4/81
0	3	1	4/81

Boxes 1	2	3	Probability
1	0	3	4/81
3	0	1	4/81
0	2	2	6/81
2	0	2	6/81
2	2	0	6/81
1	1	2	12/81
2	1	1	12/81
1	2	1	12/81

(a)

y \ x	0	1	2	3	4
0	0	24/81	12/81	0	0
1	14/81	8/81	12/81	8/81	0
2	2/81	0	0	0	1/81

(b)

x	0	1	2	3	4
$f(x)$	$\frac{16}{81}$	$\frac{32}{81}$	$\frac{24}{81}$	$\frac{8}{81}$	$\frac{1}{81}$

y	0	1	2
$g(y)$	$\frac{36}{81}$	$\frac{42}{81}$	$\frac{3}{81}$

(c) No.

16.

x	0	1	2
$f(x)$	3/14	8/14	3/14
$F(x)$	3/14	11/14	14/14

17. (a)

E_i	M	FM	FFM	FFF
$\Pr(E_i)$	p	$(1 - p)p$	$(1 - p)^2 p$	$(1 - p)^3$

(b)

g	0	1	2	3
$f(g)$	p	$(1 - p)p$	$(1 - p)^2 p$	$(1 - p)^3$

(c)

y	0	1	2	3
$h(y)$	p^3	$3(1 - p)p^2$	$3(1 - p)^2 p$	$(1 - p)^3$

(d)

$g = y$	0	1	2	3
$f(g)$	1/2	1/4	1/8	1/8
$h(y)$	1/8	3/8	3/8	1/8

19. (b)

x	4	5
$f(x)$	$\left(\frac{3}{4}\right)^4 + \left(\frac{1}{4}\right)^4$	$4\left[\left(\frac{3}{4}\right)^4\left(\frac{1}{4}\right) + \left(\frac{1}{4}\right)^4\left(\frac{3}{4}\right)\right]$

x	6	7
$f(x)$	$10\left[\left(\frac{3}{4}\right)^4\left(\frac{1}{4}\right)^2 + \left(\frac{1}{4}\right)^4\left(\frac{3}{4}\right)^2\right]$	$20\left[\left(\frac{3}{4}\right)^4\left(\frac{1}{4}\right)^3 + \left(\frac{1}{4}\right)^4\left(\frac{3}{4}\right)^3\right]$

20.

y \ x	0	1	2	$f_2(y)$
0	(.3)(1)	(.4)(.5)	(.3)(.25)	.575
1	(.3)(0)	(.4)(.5)	(.3)(.5)	.35
2	(.3)(0)	(.4)(0)	(.3)(.25)	.075
$f_1(x)$	.3	.4	.3	

X and Y are not independent.

21. (a) $S = \{(x, y) \mid x = 3, 4, 5; y = 1, 2, \ldots, x - 1\}$

(b)

y \ x	3	4	5	
1	$\left(\frac{1}{3}\right)\left(\frac{1}{2}\right)$	$\left(\frac{1}{3}\right)\left(\frac{1}{3}\right)$	$\left(\frac{1}{3}\right)\left(\frac{1}{3}\right)$	$\frac{13}{36}$
2	$\left(\frac{1}{3}\right)\left(\frac{1}{2}\right)$	$\left(\frac{1}{3}\right)\left(\frac{1}{3}\right)$	$\left(\frac{1}{3}\right)\left(\frac{1}{4}\right)$	$\frac{13}{36}$
3	0	$\left(\frac{1}{3}\right)\left(\frac{1}{3}\right)$	$\left(\frac{1}{3}\right)\left(\frac{1}{4}\right)$	$\frac{7}{36}$
4	0	0	$\left(\frac{1}{3}\right)\left(\frac{1}{4}\right)$	$\frac{3}{36}$
	$\frac{1}{3}$	$\frac{1}{3}$	$\frac{1}{3}$	

X and Y are not independent.

22.

y	0	1	3	2	4
f(y)	1/5	1/5	1/5	1/5	1/5
x	0	1	2	3	4

23. (a)

y	0	1	4
f(y)	1/5	2/5	2/5

(b)

z	0	1
f(z)	1/5	4/5

(c) $W = Z^2 \pmod 5$

w	0	1
f(w)	1/5	4/5

24. $p_1(x) = x/11$, $x = 2, 4, 5$ $p_2(y) = y/6$, $y = 1, 2, 3$;
X and Y are independent.

25.

y \ x	0	1	2	3	4
-1	1/16	4/16	0	0	0
0	0	0	6/16	0	0
1	0	0	0	4/16	1/16

X and Y are <u>not</u> independent.

Chapter 7

1. (a) $E(X) = 0$, $Var(X) = 1.6$

   (b)

y	-7	-4	-1	2	5
f(y)	.2	.1	.3	.3	.1

   $E(Y) = -1$, $Var(Y) = 14.4$

   (c) $3E(X) - 1 = -1 = E(Y)$, $9Var(X) = 9(1.6) = 14.4 = Var(Y)$

2. $E(X) = .55$, $E(Y) = 2.1$, $Var(X) = 1.3475$, $Var(Y) = 5.39$

   $E(Y) = 2E(X) + 1$, $Var(Y) = 4Var(X)$

3. (a) $Y = Z - c$ $\quad E(Z) = E(Y) + c = \mu_Y + c$

   $Var(Z) = Var(Y) = \sigma_Y^2$

   (b)

v	-2	-1	0	1	2
Pr(V = v)	.2	.2	.3	.2	.1

   $E(V) = -.2$, $E(X) = 76.8$

   $Var(V) = 1.56$, $Var(X) = 1.56$

   (c) $E(Z) = dE(Y) + c$, $Var(Z) = d^2Var(Y)$

   (d) $E(C) = 5\{E(F) - 32\}/9$, $Var(C) = (25/81)Var(F)$

   $E(F) = (9/5)E(C) + 32$, $Var(F) = (81/25)Var(C)$

4. (a) WR:

x	0	1	2	3
$f(x)$	.216	.432	.288	.064

WOR:

x	0	1	2	3
$f(x)$	.167	.500	.300	.033

(b) Expected number:

WR	21.6	43.2	28.8	6.4
WOR	16.7	50.0	30.0	3.3

(c) Observed frequencies more closely resemble sampling WOR.

5. $\sum_{i=1}^{k} (x_i - c)f(x_i) = \Sigma\, x_i f(x_i) - c\, \Sigma\, f(x_i) = E(X) - c$; hence, if $\Sigma\, (x_i - c)f(x_i) = 0$, then $c = E(X)$.

6. $E[(X - a)^2] = E\{(X - E(X) + E(X) - a)^2\}$

$$= E\{(X - E(X))^2 + 2[X - E(X)][E(X) - a] + [E(X) - a]^2\}$$

$$E\{(X - a)^2\} = E\{[X - E(X)]^2\} + E\{[E(X) - a]^2\} + 2[E(X) - a]E[X - E(X)]$$

$$= \text{Var}(X) + [E(X) - a]^2 + 0.$$

Thus, a equal to $E(X)$ minimizes $E\{(X - a)^2\}$.

7. $E(Y) = a + bE(X) + cE(X^2)$. Recall that $E(X^2) = \text{Var}(X) + [E(X)]^2$; thus

$$E(Y) = a + bE(X) + c\text{Var}(X) + c[E(X)]^2$$

10. (a) $E(Y) = \sum_{i=1}^{k} |x_i - E(X)|\, f(x_i)$

$$E(Z) = \sum_{i=1}^{k} \{x_i - E(X)\}^2 f(x_i)$$

Since "absolute value" and "squaring" always give positive quantities and $f(x_i) \geq 0$, $E(Y)$ and $E(Z)$ will be sums of positive quantities.

(b) If $\Pr[X = E(X)] = 1$, then $E(Y) = 0$, $E(Z) = 0$.

(c) E(Y) and E(Z) increase as X deviates from E(X).

(d) Variance of X

11.

	0	1	2	E(S)	Var(S)
$f_1(s)$	.32	.56	.12	.8	.4
$f_2(s)$	$(.6)^2$	2(.4)(.6)	$(.4)^2$	.8	.48
$f_3(s)$	.3588	.4824	.1588	.8	.4776
$f_4(s)$	.4	.4	.2	.8	.56
$f_5(s)$	.398	.404	.198	.8	.556

12. (a)

x	1	2	3
f(x)	p	p(1 - p)	$(1 - p)^2$

(b) $E(X) = p^2 - 3p + 3$, $Var(X) = p(5 - 10p + 6p^2 - p^3)$

15. (a) 10025, (b) 100, (c) 10, (d) -100, (e) 25, (f) 5

16. (a) $E(X) = p$, $Var(X) = p(1 - p)$

(b) $Var(X) = p(1 - p)$ has maximum of 1/4 when $p = 1/2$.

(c) $E(X^2) = E(X) = k$, then $Var(X) = k - k^2 = k(1 - k)$.

$Var(X) \leq 1/4$ for $0 \leq k \leq 1$.

(d) Indicator random variable.

17. (a) $E(X) = 3$, $Var(X) = 2$

(b)

y	1	2	3	4	5
f(y)	1/25	3/25	5/25	7/25	9/25

$E(Y) = 95/25 = 3.8$, $Var(Y) = 1.36$

(c)

y	1	2	3	4	5
f(y)	0	1/10	2/10	3/10	4/10

$E(Y) = 4$, $Var(Y) = 1.0$

18. 5287/1024 = 5.16

19. (a)

z	a	a + 1	...	a + b
f(z)	$\frac{1}{b + 1}$	$\frac{1}{b + 1}$	...	$\frac{1}{b + 1}$

or $f(z) = \frac{1}{b + 1}$ for $z = a, a + 1, \ldots, a + b$

(b) $E(Z) = \left(\frac{1}{b + 1}\right)\{a + (a + 1) + \cdots + (a + b)\} = a + b/2$

(c) $Var(Z) = \frac{b(b + 2)}{12}$

20. (a) $f(x) = \frac{1}{n}$ for $x = 1, 2, ..., n$

$F(x) = \frac{x}{n}$ for $x = 1, 2, ..., n$

(b) $E(X) = \frac{n + 1}{2}$, $Var(X) = \frac{n^2 - 1}{12}$

(c) (i) Rolling a balanced die with n = 6.

(ii) Drawing a random number from 01, 02, ..., 99 with n = 99.

(d) a = 1, b = n - 1

21. New Method:

E(profit) = .5875

	D	$\bar{D}$
Probability	.05	.95
Profit	-.60	.65

Old Method:

E(profit) = .75

	D	$\bar{D}$
Probability	.08	.92
Profit	-.40	.85

Continue with old method.

22. Assume acb is correct; then

abc	acb	bac	bca	cab	cba
1	3	0	1	1	0

E(X) = 1

23. S = {D, GD, GGD, GGG} with probabilities 1/4, 1/4, 1/4, 1/4. Since X = 3 for GGD and GGG, E(X) = 2.25.

25. (a) \$2.50, (b) \$3.33

26. (a) Jack's expected net gain = $\frac{10}{36}$ (\$2) + $\frac{1}{36}$ (\$4) - \$1 = -33¢.

(b) He should pay 67 cents if the game is to be fair.

27. (b) $E(Y_1) = 2$, $E(Y_2) = 1.8$

28. $E(G_1) = -2/3$, $Var(G_1) = 8/3$

$E(G_2) = -2/3$, $Var(G_2) = 10/3$

29. X = label on chosen ball. Then

$$f(x) = \frac{2x}{n(n + 1)} \quad \text{for } x = 1, 2, \ldots, n$$

$$E(X) = \frac{2n + 1}{3}, \quad \text{Var}(X) = \frac{(n + 2)(n - 1)}{18}$$

33. $f(t) = \dfrac{t - 1}{36}$ for $t = 2, 3, \ldots, 7$

$= \dfrac{13 - t}{36}$ for $t = 8, 9, \ldots, 12$

E(T) = Med(T) = Mode(T) = 7

37. (a) $E(X) = 0$, $\text{Var}(X) = 2k^2p$

(b) p = 1/2 with:

x	-k	0	k
f(x)	1/2	0	1/2

(c) p = 1/8, (d) p = 1/18

(e) $p = \dfrac{1}{2c^2}$, thus $\Pr\{|X - \mu| \geq c\sigma\} = \Pr\{|X| \geq k\} = \dfrac{1}{c^2}$,
that is, the equality is attained.

Chapter 8

2. (a) E(Z) = E(W) = 3/2, Var(Z) = Var(W) = 3/4

(b) Cov(Z, W) = -3/4; Z and W are not independent.

(c), (d) E(Z + W) = 3, Var(Z + W) = 0

3. (b) E(X) = 20/3, E(Y) = 15/7, E(XY) = 58/3

(c) X and Y are not independent.

4. (b) E(X + Y) = 4, Var(X + Y) = 1, E(XY) = 4, E(X/Y) = 1.2

(c) E(X) = 2, Var(X) = .4, E(Y) = 2, Var(Y) = .6

(d) $E(X + Y) = E(X) + E(Y)$, $E(XY) = E(X)E(Y)$, $E(X/Y) \neq E(X)/E(Y)$

(e) X and Y are not independent since $f(x, y) \neq f_1(x)f_2(y)$ for all (x, y). Var(X + Y) = Var(X) + Var(Y) since Cov(X, Y) = 0.

5. (a) E(X) = E(Y) = 2/3, Var(X) = Var(Y) = 5/9, Cov(X, Y) = -5/18, $\rho_{XY} = -1/2$

(b) $Var(X + Y) = 5/9$, (c) No

(d) $Var(X + Y) \neq Var(X) + Var(Y)$ since $Cov(X, Y) \neq 0$.

6. $E(W_1) = 65$, $Var(W_1) = 425/4$,
$E(W_2) = 200/3$, $Var(W_2) = 400/3$.

7. $E(W_1) = 65$, $Var(W_1) = 125/4$,
$E(W_2) - 200/3$, $Var(W_2) = 200/3$.
A negative covariance indicates that X_1 and X_2 are inversely related. The negative covariance has reduced the $Var(W_1)$ and $Var(W_2)$.

8. $E(W_1) = 65$, $Var(W_1) = 325/4$,
$E(W_2) = 200/3$, $Var(W_2) = 1000/9$.
When $Cov(X_1, X_2) = 0$, X_1 and X_2 are unrelated.

10. $E(T) = n\mu_X$, $Var(T) = n\sigma_X^2$

11. (a) 110, 325, -10, 325, (b) 0, 0, 2, 2

14. (a) $\mu_X = \dfrac{b + d}{n}$, $\mu_Y = \dfrac{c + d}{n}$,

$$\sigma_X^2 = \frac{(b + d)(a + c)}{n^2}, \quad \sigma_Y^2 = \frac{(c + d)(a + b)}{n^2}$$

(b) $\sigma_{XY} = \dfrac{ad - bc}{n^2}$

15. (a) $\rho_{XY} = 0$; X and Y are independent since $\rho = 0$ implies independence for indicator random variables.

(b) $\rho_{XY} \neq 0$; X and Y are <u>not</u> independent.

16. $\rho_{XY} = 0$

20. $Var(U) = \sigma^2(k^2 + 1) \approx k^2\sigma^2$ since k is much greater than 1. $Var(U)$ is approximately the same as $Var(Y)$; the X component is negligible.

21. $E(X^*) = 0$, $Var(X^*) = 1$, $Pr\{|X^* - \mu| \leq h\sigma\} > 1 - \dfrac{1}{h^2}$ <u>or</u>
$Pr\{|X^*| \leq h\} > 1 - \dfrac{1}{h^2}$

22. (a) $<.25$, (b) $>.75$, (c) $\sqrt{50}$

23. (a) $Var(S) = n\sigma^2[1 + (n - 1)\rho]$, (b) $\rho = -\frac{1}{n - 1}$

24. (b) $\rho(V, W) = 1/2$

25. $\rho(X, Z) = \rho(X, X + Y) = \sqrt{2}/2$

26. (a) $E(X) = 2$, $E(Y) = 6$, $E(Z) = 2$,

(b) $Var(X) = 8/5$, $Var(Y) = 4/5$, $Var(Z) = 8/5$.

X and Y are independent and so are Y and Z.

(c) $E(X + Y + Z) = 10$, $Var(X + Y + Z) = 140/25$

[Hint: See Problem 24(a) for computing Cov(X, Z).]

27. $E(X) = 14/5$, $Var(X) = 14/25$, $E(\bar{X}) = 14/5$, $Var(\bar{X}) = 14/25n$

28. (a) $\mu_X = -1/4$, $\sigma_X = \sqrt{7}/4$

(b) There are 3^2 possible samples; the p.f. of $\bar{X}$ will be:

$\bar{x}$	-1	-1/2	0	1/2	1
$f(\bar{x})$	$(3/8)^2$	2(3/8)(1/2)	11/32	2(1/2)(1/8)	$(1/8)^2$

(c) $E(\bar{X}) = -1/4$, $Var(\bar{X}) = 7/32$

29. (b) 0, (c) $\sigma^2\left(\frac{1}{n_F} + \frac{1}{n_M}\right)$

30. (a) $E(W) = \mu$, (b) $Var(W) = \frac{a^2\sigma^2}{n_1} + (1 - a)^2 \frac{\sigma^2}{n_2}$,

(c) $a = \frac{n_1}{n_1 + n_2}$, (d) $a = 1/2$

31. $n = 32$

33. (a) $\frac{1}{10}$, $\frac{1}{10}$, (b) $\frac{1}{90}$, (d) $E(T) = 1$, $Var(T) = 1$

Chapter 9

1. (a) .0170, .0989, .6553; (b) 0, .6481, .0138, (c) .8

2. $n = 10$, $p = .6$

(a) .9877, (b) .8205, (c) .9537

3. 69

4. (a) 7/27, (b) 26/27, (c) 2/9

6. (a) np, $np(1 - p)$, (b) $S \sim b(s; n, p)$, (c) .0019

7. (a) $P = S/n$, (b) p, $p(1 - p)/n$, (c) .0019

   (d) As n becomes larger, Var(P) gets smaller; that is, P will approach its true value p.

8. $n(1 - p)$, $n(1 - p)p$

9. $p = x/n$

10. (a) .0035, .7759, .2206, (b) .7759, .0035, .2206

12. $1/2^{14}$, $1/2^{14}$

13. (a) .6

No. of P	0	1	2	3	4	5	6	7	8
f	0	0	5	14	21	29	20	9	2
$E(f \mid p = .6)$	0	.8	4.1	12.4	23.2	27.9	20.9	9.0	1.7
$E(f \mid p = .7)$	0	.1	1.0	4.7	13.6	25.4	29.6	19.8	5.8

The fit is better with p = .6.

19. Pr(Felix gets three healthy birds on a single try) = 1/10; Pr(two successes in three tries) = .027.

20. (a) 1/2, (b) $b(x; 10, 1/2)$,

    (c) $Pr(X \geq 9) = .0108$. Conclude that drug slows reaction time since the probability is very small.

21. (a) $Pr(X \leq 3 \mid n = 10, p = .6) = .0548$. Newspaper A's claim is not supported since $Pr(X \leq 3)$ is small.

    (b) .6496, .3822, .1719. These probabilities indicate that the event $X \leq 3$ is far more probable if p = .6; hence, one would conclude that $p \neq .6$ as claimed by A, but actually p is less than .6.

22. (a) .1768, (b) .4963

23. (a) If each of the methods is equally effective, then the ranks of the children for method II represent a random selection of three numbers from 1, 2, 3, 4, 5, 6. Each

selection has probability $1/\binom{6}{3} = 1/20$. The sample space consists of 20 points:

E_i	T
(4, 5, 6)	15
(3, 5, 6)	14
(2, 5, 6) (3, 4, 6)	13
(1, 5, 6) (2, 4, 6) (3, 4, 5)	12
(1, 4, 6) (2, 3, 6) (2, 4, 5)	11
(1, 3, 6) (1, 4, 5) (2, 3, 5)	10
(1, 2, 6) (1, 3, 5) (2, 3, 4)	9
(1, 2, 5) (1, 3, 4)	8
(1, 2, 4)	7
(1, 2, 3)	6

t	6	7	8	9	10	11	12	13	14	15
f(t)	$\frac{1}{20}$	$\frac{1}{20}$	$\frac{2}{20}$	$\frac{3}{20}$	$\frac{3}{20}$	$\frac{3}{20}$	$\frac{3}{20}$	$\frac{2}{20}$	$\frac{1}{20}$	$\frac{1}{20}$

(b) E(T) = 10.5, Var(T) = 5.25

(c) If method II is more effective, T < 10.5 (i.e., smaller values being more convincing). The observed event T = 7 seems to indicate that method II (not I) is more effective.

24. (a) Let U indicate up and D indicate down. Then a typical point in S is (x_1, x_2, x_3, x_4) with x_i = D or U. There are $2^4 = 16$ points in S.

(b) Let C = number of changes

E_i	C
(UUUU) (DDDD)	0
(UUUD) (DUUU) (UUDD) (DDUU) (DDDU) (UDDD)	1
(UUDU) (UDUU) (UDDU) (DUUD) (DDUD) (DUDD)	2
(DUDU) (UDUD)	3

c	0	1	2	3
Pr(C = c)	2/16	6/16	6/16	2/16

$\Pr(C \geq 2) = 1/2$

(c) Let X = number of cases with two or more changes; then $X \sim b(x; n, 1/2)$. Assumption: (i) Pr(two or more changes) is same for each case, (ii) Cases are independent. (iii) Considering only two outcomes: ≥ 2 and <2 changes. (iv) Number of cases for consideration is fixed in advance.

(d) $\Pr\{X \geq 13 \mid X \sim b(x; 20, 1/2)\} = .1316$. Since $\Pr(X \geq 13)$ is reasonably large, the data support the hypothesis of randomness.

25. (a) .8542, (b) .7463

26. (a) Let X be the number of uncooked peas in a sample of 10 peas. Since the sample size is small in comparison with the population size of two pounds of peas, X is approximately $b(x; 10, p)$. Pr(declaring peas cooked) equals $\Pr(X = 0) = (1 - p)^{10}$. Hence

p	.1	.2	.3	.4	.5	.6	>.6
Pr(X = 0)	.3487	.1074	.0282	.0060	.0010	.0001	.0000

(b) .3487

(c) This sampling scheme may be too stringent. We might use X = 0 or 1 as an acceptable region.

27. (a) T is actually a hypergeometric variable, but its distribution can be approximated by a binomial distribution since the sample size 5 is small in comparison with the population size.

(b)

p	.1	.2	.3	.4	.5
Pr(T = 0)	.5905	.3277	.1681	.0778	.0313

(e) The distribution of T can no longer be approximated by a binomial since a sample of 150 from 500 is a 30% sample. The random variable T will have a hypergeometric distribution.

28. Pr(engine does not fail) = $1 - q$ and Pr(engine fails) = q; hence, Pr(two-engine plane runs) = $1 - q^2$ and Pr(four-engine plane runs) = $1 - q^4 - 4q^3(1 - q)$. The two-engine plane will be preferable when $1 - q^2 \geq 1 - q^4 - 4q^3(1 - q)$ or $q^2(3q - 1)(q - 1) \leq 0$. If $q = 0, 1/3, 1$, the performance will be equal. For $0 < q < 1/3$, four-engine plane preferred and for $1/3 < q < 1$, two-engine plane preferred. Since q would usually be much less than 1/3, the four-engine plane would more than likely be preferred.

31. (a) 3/4, (b) $E(X) = 1/3$, $Var(X) = 4/9$

32. (a) $k = 1 - r$, (b) $E(Z) = \frac{r}{1 - r}$, $Var(Z) = \frac{r}{(1 - r)^2}$

33. (a) $Pr(X = k) = (.9)^{k-1}(.1)$, $k = 1, 2, \ldots$
    (b) $Pr(X > t) = (.9)^t$
    (c) $Pr(X = j \mid X > t) = (.1)(.9)^{j-t-1}$, for $j = t + 1, t + 2, \ldots$

34. (b) 1/6, (c) $= \left(\frac{1}{2}\right)^{11}$

35. (a) $f(x) = \binom{x - 1}{1}\left(\frac{3}{4}\right)^2\left(\frac{1}{4}\right)^{x-2}$ for $x = 2, 3, \ldots$
    (c) $E(X) = 8/3$, $Var(X) = 8/9$

36. (a) (i) $\binom{19}{7}\left(\frac{1}{2}\right)^{20}$, (ii) $1 - \sum_{x=8}^{20}\binom{x - 1}{7}\left(\frac{1}{2}\right)^x$, (b) 16

38. (a) $\frac{7}{2^8}$, (b) $7\left(\frac{1}{6}\right)^2\left(\frac{5}{6}\right)^6$

39. (a) $E(X) = 2$; expected cost = \$200
    (b) Maximum probability occurs when $X = 1$ or $X = 2$; hence, most probable amount of money is \$150.
    (c) .0526

40. $X \sim p(x; 2)$; hence $Pr(X \leq 5) = .9835$ and $Pr(X \leq 4) = .9473$. Thus the merchant should have at least five items.

42. Probability of error = .02

(a) Let X_i = number of errors in a block of 10 words; $X_i \sim b(x_i; 10, .02)$. Then $T = \sum_{i=1}^{10} X_i \sim b(t; 100, .02)$. Since n is large and p is small, $T \overset{approx.}{\sim} p(t; 2)$ with $Pr(T \leq 1) \cong .4060$.

(b) $T \sim b(t; 100, .02)$ which again can be approximated by a Poisson with $\lambda = 2$. The probability is the same as in (a).

43. Let X_i = number of students opposing the new grading scheme in a sample (WOR) of n_i from ith class for i = 1, 2, 3, 4. Then $X_i \sim h(x_i; .1N_i, N_i, n_i)$ or $X_i \overset{approx.}{\sim} b(x_i; n_i, .1)$ since n_i is less than 20% sample for all i. Thus the total number opposing is $T = \Sigma X_i \overset{approx.}{\sim} b(t; \Sigma n_i, .1)$ or $T \overset{approx.}{\sim} b(t; 100, .1)$. Since Σn_i is large and p is small, the distribution of T can further be approximated by $T \overset{approx.}{\sim} p(t; 10)$ with $Pr(T \geq 10) = .5421$.

44. Let X_i = number of defectives in sample of 100 from each machine for i = A and B; $X_A \overset{approx.}{\sim} p(x; 1)$, $X_B \overset{approx.}{\sim} p(x; 2)$. Hence $S = X_A + X_B \sim p(s; 3)$ and $Pr(S > 1) = .8008$.

46. (a) Let X = number of donors with a particular blood type; $X \sim b(x; 5, .03)$.

(b) Let Y = number of defectives in sample of 25 sets; $Y \sim h(y; 400, 5000, 25)$. Since 5000 >> 25, $Y \overset{approx.}{\sim} b(y; 25, .08)$, which, since n = 25 is reasonably large and p = .08 is small, can be approximated by a Poisson with $\lambda = 2$.

(c) Let T = number of fish caught; $t \sim nb(t; 10, .3)$.

(d) Let X = number in sample produced by first shift; $X \sim h(x; 400, 1000, 5)$ or $X \overset{approx.}{\sim} b(x; 5, .4)$.

(e) Let Z = number of customers with colored telephones; $Z \sim h(z; 100, 100{,}000, 100)$ or $Z \overset{approx.}{\sim} b(z; 100, .001)$ or $Z \overset{approx.}{\sim} p(z; .1)$.

Index

N

O

P

R

S

T

U

V